Control Theory and Biological Systems

Control Theory

 1963

New York and London

and Biological Systems

FRED S. GRODINS

COLUMBIA UNIVERSITY PRESS

OPTOMETRY

Preface

ALTHOUGH physiologists have realized since the time of Claude Bernard that they deal with a collection of complex, interrelated biological regulators, the application of control-system theory to biological systems has only just begun. This is partly because the theory itself is of relatively recent origin and was first developed for physical systems by mathematicians, physicists, and engineers whose language is often unfamiliar to biologists. This communication barrier has served to conceal the beautiful generality of systems theory. It does not matter in principle whether the system is a physical one built by man or whether it is that most complex and marvelous system of all, man himself. This does not mean that all of the theoretical tools necessary for the analysis of biological systems are already available, for biological systems are much more complex than physical systems. Thus, existing control-system theory is largely a theory of linear systems but almost all biological systems contain essential nonlinearities. Although a complete analytical theory exists for linear differential equations, no comparable general treatment is yet available for nonlinear ones. However, for particular cases, solutions can now be obtained by computer techniques.

But the important point is that whatever form the equations may take for a particular system, the general approach is the same for all systems. This is the theme that I will emphasize in this book. The very act of formulating the problem in terms of block diagrams and mathematical relationships, requiring as it does the precise identification and rigorous definition of previously vague concepts, provides insight and clarification obtainable in no other way. In this sense, the details of the mathematical manipulations required to solve the particular equations so obtained are of secondary importance.

We would be unrealistic if we did not recognize a traditional reluctance to employ mathematical methods in biology. Although biologists do not deny the value of mathematical theories in the physical sciences, they often seem to forget that such theories deal almost entirely with nonexistent abstractions, e.g., ideal gases, ideal solutions, perfectly elastic particles, massless springs, weightless frictionless pistons, etc., ad infinitum. In any case, attempts to employ similar methods in the analysis of biological systems often encounter vigorous opposition. Perhaps this attitude stems from a misunderstanding of the purpose of abstract theory. Such theories were never meant to describe in minute detail every conceivable aspect of system structure and function. Instead they are intended to provide a useful means of thinking about and of "understanding" the system. That this kind of understanding eventually leads to practically useful results is evidenced by much of our modern technology.

In the chapters to follow, I hope to demonstrate the power and the usefulness of the systems approach in advancing our understanding of complex biological regulators. To provide the necessary background for the study of biological systems, we shall begin with a discussion of systems in general and follow this with a consideration of particular physical systems. Finally, we shall consider two examples of biological control systems, the respiratory and the cardiovascular.

It would be impossible to adequately thank everyone who has helped make this book possible. The author is particularly grateful to Professor John S. Gray who introduced him to and stimulated his study of biological control systems, and to Professor Richard W. Jones who introduced him to control-system theory. Constant reference was made to a number of texts in preparing the basic material of the first five chapters. Perhaps the one most generally useful and influential was John D. Trimmer's *Response of Physical Systems* (New York, John Wiley and Sons, Inc., 1950).

Other useful texts whose inadequately acknowledged contributions should be frequently recognizable are: M. F. Gardner and J. L. Barnes, *Transients in Linear Systems* (New York, John Wiley and Sons, Inc., 1942); J. A. Greenwood, Jr., J. V. Holdam, Jr., and D. Macrae, Jr., eds., *Electronic Instruments*

(New York, McGraw-Hill Book Company, Inc., 1948, Chapters 8–11); G. S. Brown and D. P. Campbell, *Principles of Servomechanisms* (New York, John Wiley and Sons, Inc., 1948); C. R. Wylie, Jr., *Advanced Engineering Mathematics* (New York, McGraw-Hill Book Company, Inc., 1951); G. H. Farrington, *Fundamentals of Automatic Control* (New York, John Wiley and Sons, Inc., 1951); J. G. Truxal, *Automatic Feedback Control System Synthesis* (New York, McGraw-Hill Book Company, Inc., 1955); G. J. Murphy, *Basic Automatic Control Theory* (Princeton, N. J., D. Van Nostrand Company, Inc., 1957); G. K. Tucker and D. M. Wills, *A Simplified Technique of Control System Engineering* (Philadelphia, Minneapolis-Honeywell Regulator Company, 1958); E. Mishkin and L. Braun, Jr., eds., *Adaptive Control Systems* (New York, McGraw-Hill Book Company, Inc., 1961); and G. J. Thaler and M. P. Pastel, *Analysis and Design of Nonlinear Feedback Control Systems* (New York, McGraw-Hill Book Company, Inc., 1962).

Several graduate students made major contributions to the studies described in Chapters 7 and 8, and particular thanks are due to Karl Schroeder, Art Norins, Bill McAdam, Gordon James, Loren Anderson, Diana Zingher, and Steve Propeck. Generous financial support for these studies has been continuously provided by the National Institutes of Health (Research Grant H-1626 and Research Career Award HE-K6-14187 from the National Heart Institute). The author is indebted to another graduate student, John Buoncristiani, for his careful critical review of the manuscript which revealed a number of errors. Finally, the author is grateful to the staff of Columbia University Press for their constant encouragement, generous technical help, and infinite patience.

Northwestern University Medical School   FRED S. GRODINS
Chicago
April 10, 1963

Contents

Control Theory and Biological Systems

CHAPTER 1

Systems in General

Basic ingredients of a system

THE system concept is a very general one. We may define a system as a collection of components arranged and interconnected in a definite way. The components may be physical, chemical, biological, or a combination of all three. If physical, they may be mechanical, electrical, thermal, etc. A distinguishing feature of such a system is that it has an input and an output. More colorful terms for the former include cause, stimulus, driving function, forcing function, etc., and, for the latter, effect, response, reaction, etc.

These terms correctly imply that if the input is varied in a particular way, then the output will be caused to vary in a particular way. The dependence of the output upon the input is defined by the law of the system. Ideally, this law can be expressed in the form of a mathematical equation having a general analytical solution. Such an equation will include a

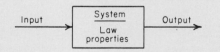

FIG. 1. GENERAL SYSTEM BLOCK DIAGRAM

number of constants which characterize the properties of the system. This basic system description may be conveniently expressed in the form of a block diagram (Fig. 1).

A distinction of considerable importance in all systems work is that between a transient and a steady-state output. If the input is suddenly changed from one constant value to another or from one periodic function to another, the output will go through a temporary adjustment (or transient) period before finally settling down to a new steady pattern. The two common techniques of system analysis, transient analysis and frequency analysis, are concerned, respectively, with the transient response to a step-function input (i.e., a sudden change from one constant input to another), and with the steady-state response to a sinusoidal input.

Problems posed by systems*

Various sorts of problems arise in connection with the basic system of Fig. 1. For example, we might be given the input, the law, and the properties and asked to predict the output. This "direct" problem is perhaps the simplest, and certainly the most familiar one, for it has long been the usual form of didactic presentations of science. Its solution is frequently required in engineering applications. On the other hand, we might be given the law, properties, and output and asked to determine the input. This "converse" problem is one sort of diagnostic problem which might be encountered by a physician. Again, we might be given the input, law, and output and asked to determine values for the system properties. This "inverse" problem is another sort of diagnostic problem encountered by the physician, for example, in interpreting pulmonary function tests. It is also encountered by the engineer as an aspect of the "design" problem in which he must determine values for the system properties which will meet certain performance specifications. Finally, we might be given only the input and the output and asked to determine the law and properties of the system. This inductive or "black box" problem is the most difficult of all, and provides the basis for the solution of all of the others. It is the problem faced by the scientist.

Thus it becomes apparent that although the engineer, the practicing physician, and the scientist all deal with systems,

* J. D. Trimmer, *Response of Physical Systems* (New York, John Wiley and Sons, Inc., 1950).

they generally face different aspects of the system problem. The scientist is initially confronted with the inductive problem. He must find laws to define the operation of systems. To do this, he isolates a system as best he can, applies known inputs to it, and observes the output. The first laws which he formulates may simply be empirical descriptions cast in mathematical form relating particular outputs to particular inputs. These empirical equations will contain constants which may have no particular theoretical or general significance.

Eventually, however, the scientist tries to obtain a law based on theoretical assumptions. The advantage of such a law is its generality, i.e., it will successfully predict the output for *any* input and for *any* values for the system properties. How does he formulate and test such a law? Its formulation involves an intangible process which might be called educated guessing. The observed behavior of the system somehow suggests to him a law. To test it, he proceeds to solve the direct (or deductive) problem and compares prediction with observation.

One of the most annoying problems facing the biological scientist is the difficulty of system isolation. He may not always be able to control all of the inputs to his system, nor can he always be sure that the properties of his system or even its components remain constant at all times. Hence the laws of biological systems are often statistical in nature.

The practicing physician certainly faces both the converse and inverse problems. However, he also faces a problem not yet stated. For example, in the inverse problem, the physician may conclude that certain property values have been altered but, having done so, he also wants to know what has altered them. We can say that he must determine the parametric inputs. We shall find that parametric "forcing" occurs frequently in biological systems.

The engineer generally deals with the direct problem and the design problem, i.e., the problem of designing a system to meet certain performance specifications. We have already noted that the latter involves the inverse problem, but does it not also involve the inductive problem faced by the scientist? Despite some superficial similarities, the answer to this question must be no. In solving the design problem, the engineer must synthesize a system which does not yet exist, and so, in a sense, the law of this system is unknown. However, in solving this problem, he combines a number of components whose laws he knows and adjusts their properties until the desired performance is obtained. When he has finished, he knows the law and properties of the final system because he built it. This is quite different from the task of the scientist who must deal with an existing system built of mysterious components by "a party or parties unknown."

Using this general background, let us begin our study with physical systems. We introduce them in Chapter 2.

Introduction to Physical Systems

WE begin our study with physical systems because they are the simplest ones available. By simple, we mean that they are comparatively easy to isolate and that they are linear. The significance of the former is obvious; the significance of the latter will become apparent as we proceed. Let us introduce these systems by examining some particular examples. We shall be concerned with the equations of motion of certain mechanical systems.

A zero-order system

Suppose we had the system illustrated by the "hardware" diagram of Fig. 2. A spring of stiffness K is attached to a rigid

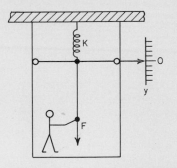

FIG. 2. ZERO-ORDER MECHANICAL SYSTEM

support above and constrained by frictionless guides to move only in the vertical direction. The man below applies a force F to the spring, thus stretching it by an amount y measured on the scale at the right. When no force is applied, the spring is at its resting position corresponding to $y = 0$. Although for simplicity we shall usually use the symbols F and y rather than $F(t)$ and $y(t)$, we will understand that in general both F and y will be functions of time t. In system terms, our input is F, our output is y, and our problem is to obtain the system law which relates them.

To obtain this equation of motion, we apply a fundamental principle of mechanics which states that at each instant of time the applied force is equal in magnitude and opposite in direction to the sum of the opposing forces generated by the motion. In the present case, our system has only one component, the spring, and this will generate the only opposing force, F_K. According to Hooke's law, F_K will be directly proportional to the displacement y:

$$F_K = Ky, \tag{2.1}$$

where K is the spring stiffness, our system property. Hence the equation of motion which provides our system law is:

$$Ky = F \tag{2.2}$$

or, solving for y,

$$y = \left[\frac{1}{K}\right] F. \tag{2.3}$$

Now Eq. (2.3) is a good starting point for our study because, being an algebraic equation, it presents no exotic problems. It says that for any given value of the input, say F_1, there will be

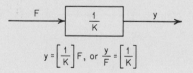

$$y = \left[\frac{1}{K}\right] F, \text{ or } \frac{y}{F} = \left[\frac{1}{K}\right]$$

FIG. 3. BLOCK DIAGRAM OF ZERO-ORDER SYSTEM

a single corresponding value of the output y_1, given by the product $[1/K]F_1$. Let us call the term $[1/K]$ the system transfer function which we define as the quantity by which the system block multiplies the input to generate the output or, alternatively, as the ratio of output to input (Fig. 3).

Let us now examine the behavior of y for a particular input F. Let this input be a step function defined by the requirements that $F = 0$ for $t < 0$ and $F = F_1$ for $t \geqq 0$. Both input and output are plotted as functions of time in Fig. 4. We note that y follows F instantaneously with no delay or lag. Hence the

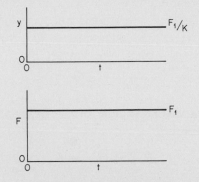

FIG. 4. STEP-FUNCTION FORCING AND RESPONSE IN
ZERO-ORDER SYSTEM

value of y depends only on the value of F and not at all on time. We call such a system, described by an algebraic equation, a zero-order system. Thus the plot of Eq. (2.3) in Fig. 5 holds for all instants of time regardless of the form of $F(t)$. If we ignore the difference in the physical units of F and y and simply regard

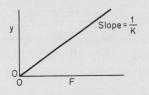

FIG. 5. GAIN CURVE

them as "signals" entering and leaving the block, we can say that the block changes the size of the input signal (i.e., it multiplies the input by a constant) but it does not change its timing or shape. To emphasize this size-changing operation, we shall call Fig. 5 a gain curve with the understanding that the gain may be either greater or less than one.

A first-order system

Let us now add a complication. Without changing the hardware diagram of Fig. 2, we shall simply say that instead of being frictionless our lubricated guides manifest viscous friction, i.e., they provide a "viscous dashpot." This means that a second opposing force F_R, proportional to velocity (i.e., to the first time derivative of displacement, dy/dt), must be added to that generated by the spring. We have:

$$F_R = R\frac{dy}{dt}, \tag{2.4}$$

where R is the viscous resistance of the dashpot. Our equation of motion thus becomes:

$$R\frac{dy}{dt} + Ky = F \tag{2.5}$$

and we have two system properties, R and K. We now enter a new world of equations.

Equation (2.5) is a differential equation, so called because it contains the derivative dy/dt. It is a first-order differential equation because the highest derivative term is a first derivative. Accordingly, we call the system which it describes a first-order system. A solution of (2.5) will consist of a functional relationship, free of derivatives, between the dependent variable y and the independent variable t. Thus the system input and output will be implicitly related through the common independent variable t. We shall postpone consideration of how we get such solutions until Chapter 3. For the present, we shall assume that we have such solutions available and can examine some of their characteristics.

Before doing this, however, let us ask whether we can put Eq. (2.5) into a form similar to that of (2.3) in the sense that the output y is expressed as the product of a transfer function and an input. The answer is yes, provided we introduce a change of symbolism in (2.5). Let us use the symbol s to indicate the operation of differentiation, i.e., let $s(y) \equiv dy/dt$. Then (2.5) can be written as:

$$(Rs +.K)y = F. \tag{2.6}$$

It proves convenient to combine the two system properties R and K into a single relational parameter, the time constant τ, defined as R/K. In these terms, (2.6) becomes:

$$(\tau s + 1)y = \left[\frac{1}{K}\right] F \tag{2.7}$$

and solving for y:

$$y = \left[\frac{1/K}{\tau s + 1}\right] F. \tag{2.8}$$

Once more we can call the bracket term the system transfer function, for it is the quantity by which the system block multiplies the input to generate the output (Fig. 6). Of course, at

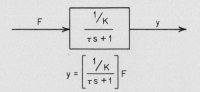

$$y = \left[\frac{1/K}{\tau s + 1}\right] F$$

FIG. 6. BLOCK DIAGRAM OF FIRST-ORDER SYSTEM

this stage, we haven't really accomplished much in solving (2.5) because so far our use of s has been merely a notational "trick," and we don't know just what it means to multiply F by a function of s. However, we will discover in Chapter 3 that when this trick is rigorously formalized in the Laplace transform method, we will encounter exactly the same form of transfer function and it will solve our differential equation for us. Hence it is helpful to get used to such transfer functions at an early stage.

Now let us examine the behavior of y for a particular input, i.e., a solution of (2.5) for a particular $F(t)$. We will again select the same step-function input used before, i.e., $F = 0$ for $t < 0$, $F = F_1$ for $t \geqq 0$, and our system will start from rest (i.e., $y = 0$) as before. In Fig. 7, we have plotted the input step function and the corresponding outputs for two different values of the time constant τ.

We at once note a difference in the behavior of this system compared to that of our zero-order example. The output y no longer follows F instantaneously, instead there is a delay or lag before y reaches its final steady-state value F_1/K. Hence the value of y following the application of a step-function input depends not only on the value of F but also on the time at which it is measured.

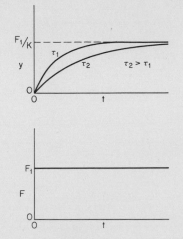

FIG. 7. *STEP-FUNCTION FORCING AND RESPONSE IN FIRST-ORDER SYSTEM*

Our system block now not only changes the size of the input signal but alters its timing or shape as well. If we used step-function inputs of various magnitudes, waited until y reached its final steady-state value for each, and then plotted this value of y against the corresponding F, we would obtain the same gain curve shown in Fig. 5. Now, however, we must call it steady-state gain, for it only holds in the steady state. Thus the constant term of the transfer function $1/K$ determines steady-state gain but says nothing about delay or lag. The other factor of the transfer function, $[1/(\tau s + 1)]$, determines the time behavior and this particular form is called a first-order lag. The output approaches its final steady-state value exponentially and shows neither overshoot nor oscillation.

A second-order system

Let us add a final complication. We shall hang a mass of inertance M on the spring and readjust our y scale so that it reads zero when only the force of gravity is acting (Fig. 8).

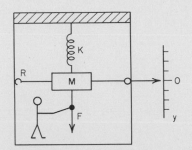

FIG. 8. *SECOND-ORDER MECHANICAL SYSTEM*

Now a third opposing force F_M will be generated by the inertial reactance of the mass. According to Newton's second law, this force is proportional to acceleration, i.e., to the second time derivative of displacement, d^2y/dt^2. We have:

$$F_M = M \frac{d^2y}{dt^2},\qquad (2.9)$$

where M is the inertance of the mass. Adding this opposing force to our other two, our equation of motion becomes:

$$M \frac{d^2y}{dt^2} + R \frac{dy}{dt} + Ky = F.\qquad (2.10)$$

Equation (2.10) is a second-order differential equation and the system which it describes is called a second-order system.

Once more it proves convenient to combine the three system properties M, R, and K, into two relational parameters defined as follows:

$$\text{Natural angular frequency} \equiv \omega_n \equiv \left(\frac{K}{M}\right)^{1/2}\qquad (2.11)$$

$$\text{Damping ratio} \equiv \zeta \equiv \frac{R}{2(KM)^{1/2}}.\qquad (2.12)$$

In terms of these parameters, (2.10) becomes:

$$\frac{1}{\omega_n{}^2}\frac{d^2y}{dt^2} + \frac{2\zeta}{\omega_n}\frac{dy}{dt} + y = \left[\frac{1}{K}\right]F. \qquad (2.13)$$

Let us again employ our notational trick to get a transfer function. If we let $s(y) \equiv dy/dt$ and $s^2(y) \equiv d^2y/dt^2$, (2.13) becomes:

$$\left(\frac{1}{\omega_n{}^2}s^2 + \frac{2\zeta}{\omega_n}s + 1\right)y = \left[\frac{1}{K}\right]F \qquad (2.14)$$

and solving for y:

$$y = \left[\frac{1/K}{(1/\omega_n{}^2)s^2 + (2\zeta/\omega_n)s + 1}\right]F. \qquad (2.15)$$

Once more the bracket term is the transfer function by which the system block multiplies the input to generate the output

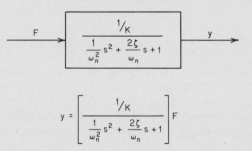

FIG. 9. BLOCK DIAGRAM OF SECOND-ORDER SYSTEM

(Fig. 9). It again contains the constant term $1/K$ which determines steady-state gain, and a function of s

$$\left[\frac{1}{(1/\omega_n{}^2)s^2 + (2\zeta/\omega_n)s + 1}\right]$$

which determines the time behavior. We call this particular form of time behavior second-order lag.

Let us now examine the response of this system to our same step-function input. Its behavior is illustrated in Fig. 10 for three different values of the damping ratio ζ. Once more we note that y does not follow F instantaneously but reaches its steady-state value F_1/K only after a lag. Moreover, something new has been added, for when $0 < \zeta < 1$, y oscillates around its final value before settling down. The occurrence of this sort of oscillation requires the presence of two energy-storage elements which can pass energy back and forth between them. In this case they are the spring, a potential-energy store, and the

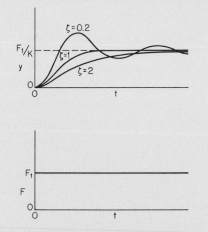

*FIG. 10. STEP-FUNCTION FORCING AND RESPONSE IN
SECOND-ORDER SYSTEM*

mass, a kinetic-energy store. If we wait for the steady state, then the value of y can be read from the gain curve of Fig. 5 as before, but, as in our first-order system, this relationship does not hold in the transient period preceding the steady state. Once more, our system block alters both the size and timing of the input signal.

Higher-order systems

By introducing additional components connected in various ways, the order of the differential equation defining our system law will increase progressively. However, as long as our system remains linear, we can always express its output in terms of linear combinations of a number of first- and second-order lags. Hence we will not consider such systems further at this time.

Analogous systems

So far we have formulated equations to describe only one particular type of physical system, i.e., a translational mechanical system. Suppose now that we examine a different sort of physical system, say an electrical circuit. Do we have to start all over again and obtain completely different forms of equations, or can we transfer what we have learned about one system to other systems? Fortunately it turns out that many systems which differ completely in their physical hardware obey exactly the same form of equation. We call such systems analogous systems. Their existence is a great convenience for it permits us to describe the behavior of many different physical systems in terms of a single mathematical model.

The analogous behavior of physical systems comprising very different hardware will perhaps seem less strange when we realize that all such systems have a common "currency," i.e., energy. This energy can always be expressed as the product of an intensity factor and a quantity factor. In our mechanical system, the intensity factor is force and the quantity factor is displacement; in an electrical system, the corresponding quantities are voltage and charge.

If a force is applied as input to a mechanical system, the system undergoes a displacement, i.e., a change of position. As a result of the motion, opposing forces are developed which are proportional to the displacement itself in a spring, to the rate of change of displacement (i.e., velocity) in a viscous dashpot, and to the rate of change of velocity (i.e., acceleration) in a mass. If a voltage is applied as input to an electrical system, the system undergoes a change in charge. As a result, opposing voltages are developed which are proportional to the charge itself in a capacitor, to the rate of change of charge (i.e., current) in a resistor, and to the rate of change of current in an inductor.

Applying Newton's second law (or d'Alembert's principle) to the mechanical system, we equate the applied force to the sum of the opposing forces; applying Kirchhoff's voltage law to the electrical system, we equate the applied voltage to the sum of the opposing voltages. Both these laws are special cases of the more general concepts of equilibrium and continuity. Hence it is not surprising that the equations describing such systems often have exactly the same form and differ only in the names assigned to the variables and constants.

To make this a bit more specific, let us examine the electrical analog of the second-order mechanical system of Fig. 8. The appropriate system is the simple *LRC* circuit of Fig. 11, con-

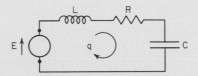

FIG. 11. SECOND-ORDER ELECTRICAL SYSTEM

sisting of an inductor (with inductance L), a resistor (with resistance R), and a capacitor (with capacitance C) connected in series. The system input is furnished by the voltage source E and the system output is the charge q on the capacitor. The system equation is obtained very simply by equating applied and opposing voltages in accordance with Kirchhoff's voltage law. The result is:

$$L \frac{d^2q}{dt^2} + R \frac{dq}{dt} + \frac{1}{C} q = E. \qquad (2.16)$$

It is at once apparent that except for the particular symbols used and the names assigned to them, Eq. (2.16) is identical with Eq. (2.10). Thus the mechanical system of Fig. 8 and the electrical system of Fig. 11 have a common mathematical model and are therefore analogous.

Analogies between the variables and properties of the two systems are summarized in Table 1. It is convenient to distinguish between active and passive system components. Active elements are energy sources. In both systems that we have considered, the input provided the active element, a force source in our mechanical example and a voltage source in the electrical example. Passive system components serve either as energy storages or as energy sinks. An energy storage may store either potential energy or kinetic energy. An energy sink converts mechanical or electrical energy into heat. In each case, the relevant property of the passive element is defined by a ratio whose numerator is the energy-intensity factor and whose de-

TABLE 1. ANALOGOUS SYSTEMS

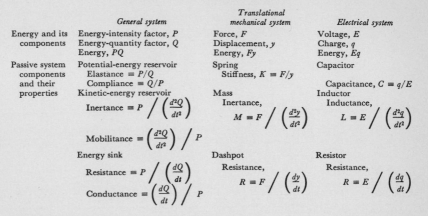

	General system	Translational mechanical system	Electrical system
Energy and its components	Energy-intensity factor, P Energy-quantity factor, Q Energy, PQ	Force, F Displacement, y Energy, Fy	Voltage, E Charge, q Energy, Eq
Passive system components and their properties	Potential-energy reservoir Elastance $\equiv P/Q$ Compliance $\equiv Q/P$	Spring Stiffness, $K \equiv F/y$	Capacitor Capacitance, $C \equiv q/E$
	Kinetic-energy reservoir Inertance $\equiv P\Big/\left(\dfrac{d^2Q}{dt^2}\right)$ Mobilitance $\equiv \left(\dfrac{d^2Q}{dt^2}\right)\Big/P$	Mass Inertance, $M \equiv F\Big/\left(\dfrac{d^2y}{dt^2}\right)$	Inductor Inductance, $L \equiv E\Big/\left(\dfrac{d^2q}{dt^2}\right)$
	Energy sink Resistance $\equiv P\Big/\left(\dfrac{dQ}{dt}\right)$ Conductance $\equiv \left(\dfrac{dQ}{dt}\right)\Big/P$	Dashpot Resistance, $R \equiv F\Big/\left(\dfrac{dy}{dt}\right)$	Resistor Resistance, $R \equiv E\Big/\left(\dfrac{dq}{dt}\right)$

nominator is the energy-quantity factor or its first or second time derivative. The reciprocal of this ratio is also sometimes used, and, in the case of the electrical capacitor, this is the usual custom.

A word of caution is perhaps indicated here. Although it was relatively easy to establish the fact that Figs. 8 and 11 represent analogous systems, it is not always so simple. Moreover, there may be more than one electrical analog of a given mechanical system. In any case, the proof of analogy always rests upon the demonstration that the systems obey the same form of equation.

Feedback systems

Since our main concern in this monograph is with control systems and since feedback is an essential feature of such systems, let us take an introductory look at the general nature and usefulness of feedback. Once more we will begin with a specific example.

Suppose that in our first-order system we provided a second

pointer y_i, not connected with the system, which could be moved along the y scale in any arbitrary fashion (Fig. 12). Then suppose we instructed our little man (the force source) to attempt to keep his pointer y_o in line with y_i at all times. What sort of block diagram can we use to describe this system? Our human

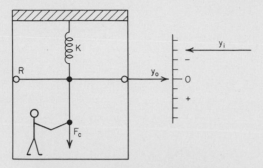

FIG. 12. A TRACKING PROBLEM

operator now visually receives two input signals: y_i, the desired value or command signal, and y_o, the actual value of the output or controlled signal. His job is to compare them and then apply a force which will move y_o to correct the "error." To keep our system a purely physical one rather than a mixed biophysical one, let us substitute an appropriate physical device (whose detailed hardware we need not specify) for the human operator in Fig. 12. The system just described can then be represented by the block diagram of Fig. 13.

We see that there are now two major components: a controlling system and a controlled system. The former includes two subcomponents. The first of these is an error detector which subtracts the output signal y_o from the command signal y_i to generate an error signal, y_e (i.e., $y_e = y_i - y_o$). The output signal required for this operation reaches the error detector via the feedback loop, and, since this signal is subtracted from y_i, the feedback is negative. The second subcomponent of the controlling system is a controller whose input is the error signal and whose output is the controlling signal, in this case the force F_c. The relationship between F_c and y_e (i.e., the controller transfer function) may be one of several kinds, but let us choose the simplest. This is proportional control in which F_c is directly proportional to y_e, i.e., $F_c = ky_e$, where k is a constant, the controller gain. The controlled system is the familiar first-order physical system already represented in the block diagram of Fig. 6. Note, however, that in Fig. 13 we have provided this

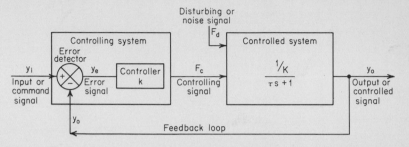

FIG. 13. BLOCK DIAGRAM OF A FEEDBACK SYSTEM

system with two inputs. One is F_c, the controlling signal. The other, F_d, we have called a disturbing or noise signal. Its significance will presently become apparent. The output of the controlled system, y_o, is fed back to the error detector. Thus the block diagram of Fig. 13 is an example of a feedback or closed-loop control system.

Our description implies that this system has a "purpose," namely, to keep the output signal y_o equal to or at least close to the command signal y_i at all times. How does feedback help accomplish this purpose? If the controller could be sure that no input other than the controlling signal F_c entered the controlled system and that the properties of the latter remained exactly constant, then it would implicitly "know" the results of its controlling operations on y_o and would not require a feedback loop to provide this information explicitly. But these are very restrictive conditions.

In general there will be other input signals acting on the controlled system, and its properties may not remain exactly constant as wear occurs and environmental conditions change. In fact, the major job of a control system may be to keep y_o constant, i.e., equal to or close to a constant command signal or set point in the face of such disturbances. The only way the controller knows that such disturbances exist is through their effect on y_o reported via the feedback loop. Hence feedback or closed-loop control has much more general application than open-loop control. It is to provide for this generality that we have included a disturbing signal F_d in Fig. 13. If the major purpose of a control system is to cause the output to follow a varying command signal, it is called a servo system; if the major purpose is to keep the output equal to or close to a constant command signal or set point, it is called a regulator. An example of the former might be an antiaircraft fire control system, and of the latter a constant-temperature water bath.

The mathematical relationship between y_o and y_i for the system of Fig. 13 is readily obtained by combining the equations for controlled and controlling systems. The former is:

$$\tau \frac{dy_o}{dt} + y_o = \frac{1}{K}(F_c + F_d) \tag{2.17}$$

and the latter:

$$F_c = k(y_i - y_o). \tag{2.18}$$

If we define $F_c/K \equiv y_c$ and $F_d/K \equiv y_d$, we can write (2.17) more conveniently as:

$$\tau \frac{dy_o}{dt} + y_o = y_c + y_d \tag{2.19}$$

and, if we define $k_1 \equiv k/K$, we can write (2.18) as:

$$y_c = k_1(y_i - y_o). \tag{2.20}$$

Substituting (2.20) into (2.19), collecting terms in y_o, and dividing through by $(1 + k_1)$ yields the desired relationship:

$$\left(\frac{\tau}{1 + k_1}\right)\frac{dy_o}{dt} + y_o = \left(\frac{k_1}{1 + k_1}\right)y_i + \left(\frac{1}{1 + k_1}\right)y_d. \tag{2.21}$$

Let us note several features of Eq. (2.21). First, like (2.17), it is a first-order differential equation. Thus the introduction of closed-loop proportional control has not altered the basic form of the input-output relation. Second, we note that the introduction of the control loop has shortened the time constant and, if k_1 (i.e., controller gain) is large, this effect will be an appreciable one. Finally, we note that a proportional control system is characterized by steady-state error.

Let us illustrate this separately for pure servo operation and for pure regulator operation. In the former, we may take $y_d = 0$. Then, if y_i is a step function, we see that the steady-state value of y_o will not be y_i but rather $[k_1/(1 + k_1)]y_i$, so that except for $y_i = 0$ the system will show steady-state error

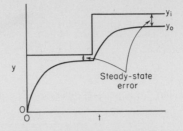

FIG. 14. SERVO OPERATION

(Fig. 14). For pure regulator operation, we may take $y_i = 0$ as the set point. Then, if y_d is a step function, the steady-state value of y_o will not be zero but $[1/(1 + k_1)]y_d$ (Fig. 15). Thus the regulator will have a steady-state error in the presence of a disturbing signal, but this error will be only $[1/(1 + k_1)]$ as

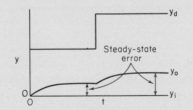

FIG. 15. REGULATOR OPERATION

large as it would be if the control loop were not present. We note that in both cases the error is minimized by making k_1 large. It can be completely eliminated by adding another form of control (integral control), but we shall postpone consideration of this to a later chapter.

The transfer function representation turns out to be a very convenient one for characterizing linear feedback systems. To illustrate this, let us recast the block diagram of Fig. 13 and the corresponding equations into this form. We first rewrite the controlled-system equation (2.19) as follows:

$$(\tau s + 1)y_o = y_c + y_d \tag{2.22}$$

and, if we now define $G(s) \equiv 1/(\tau s + 1)$, the equation assumes the following simple form:

$$y_o = G(s)[y_c + y_d], \tag{2.23}$$

where $G(s)$ is the transfer function of the controlled system. We recall that the controlling-system equation is:

$$y_c = k_1 y_e = k_1(y_i - y_o), \tag{2.24}$$

where k_1 is the controlling-system transfer function.

The block diagram of Fig. 13, modified for this notation, appears in Fig. 16. Equations (2.23) and (2.24) can now be

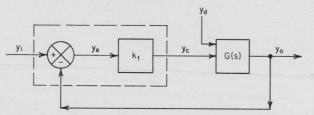

FIG. 16. SIMPLIFIED FEEDBACK SYSTEM DIAGRAM

combined in various ways to yield a number of useful relationships. In so doing, the operational symbol (s) can be treated as if it were an ordinary algebraic quantity so that the only manipulations required are algebraic ones. Thus, if we substitute $k_1 y_e$ for y_c in (2.23), we can express y_o as a function of y_e and y_d:

$$y_o = k_1 G(s)y_e + G(s)y_d. \tag{2.25}$$

If we substitute $(y_i - y_o)$ for y_e in (2.25) we get

$$y_o = k_1 G(s)[y_i - y_o] + G(s)y_d,$$

which on collecting terms in y_o and dividing through by $1 + k_1 G(s)$ yields the following expression for y_o as a function of y_i and y_d:

$$y_o = \frac{k_1 G(s)}{1 + k_1 G(s)} y_i + \frac{G(s)}{1 + k_1 G(s)} y_d. \tag{2.26}$$

Finally, since $y_e = y_i - y_o$, it is clear that $y_o = y_i - y_e$. Making this substitution for y_o in (2.26) and solving for y_e yields an expression for y_e as a function of y_i and y_d:

$$y_e = \frac{1}{1 + k_1 G(s)} y_i - \frac{G(s)}{1 + k_1 G(s)} y_d. \qquad (2.27)$$

Equations (2.25), (2.26), and (2.27) define certain transfer functions which are widely used in control-system theory.

First, taking $y_d = 0$ in (2.25), (2.26), and (2.27), the second term on the right drops out in each. Dividing the remaining expressions through by y_e [Eq. (2.25)] or by y_i [Eqs. (2.26) and (2.27)] then yields three interrelated transfer functions which characterize pure servo operation. These are:

the open-loop transfer function from (2.25)

$$\frac{y_o}{y_e} = k_1 G(s) \qquad (2.28)$$

the closed-loop output transfer function from (2.26)

$$\frac{y_o}{y_i} = \frac{k_1 G(s)}{1 + k_1 G(s)} \qquad (2.29)$$

the closed-loop error transfer function from (2.27)

$$\frac{y_e}{y_i} = \frac{1}{1 + k_1 G(s)}. \qquad (2.30)$$

There is nothing mysterious about these relationships. It is obvious from the block diagram that (2.28) is equally valid whether the feedback loop is open or closed, but that (2.29) and (2.30) hold only for the closed loop. Equation (2.28) implies that if $y_e = 1$, then $y_o = k_1 G(s)$. But if $y_e = 1$ and $y_o = k_1 G(s)$ with the feedback loop closed, then y_i must equal $1 + k_1 G(s)$ since $y_i = y_e + y_o$. Hence it is clear that with the loop closed the ratios y_o/y_i and y_e/y_i are defined by (2.29) and (2.30), respectively. If we now take $y_e = 0$ in (2.25) and $y_i = 0$ in (2.26) and (2.27), we obtain the three corresponding transfer functions for pure regulator operation:

$$\frac{y_o}{y_d} = G(s) \qquad \text{(open loop)} \qquad (2.31)$$

$$\frac{y_o}{y_d} = \frac{G(s)}{1 + k_1 G(s)} \qquad \text{(closed-loop output)} \qquad (2.32)$$

$$\frac{y_e}{y_d} = \frac{-G(s)}{1 + k_1 G(s)} \qquad \text{(closed-loop error)}. \qquad (2.33)$$

Note that in the steady state, when $G(s) = 1$, Eqs. (2.29) and (2.32) define the same errors previously described in connection with Eq. (2.21).

When studying a pure servo system, the control engineer first attempts to determine the open-loop transfer function y_o/y_e, for once this is known the closed-loop functions y_o/y_i and y_e/y_i can be written down at once. Note that for this purpose the controlling- and controlled-system blocks (excepting the error-detector function of the former) can be lumped into a single block with transfer function $k_1 G(s)$ (Fig. 17). Evidently, more detailed in-

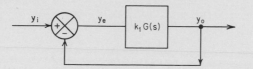

FIG. 17. SINGLE-LOOP UNITY FEEDBACK SYSTEM

formation is required to characterize the same system operating as a regulator, for here we must know not only the lumped transfer function $k_1 G(s)$ but its individual components as well.

Note that a bit of subterfuge was used to enable us to employ this simple transfer-function description, i.e., the system can operate either as a pure servo-system with $y_d = 0$, or as a pure regulator with $y_i = 0$, but it is not allowed to perform both servo and regulator functions simultaneously! If it were, then no single transfer function would serve to describe it, and we would have to return to the more general description embodied in Eqs. (2.25)–(2.27). The block-diagram transfer-function approach still remains useful, however, for we can regard the closed-loop output y_o in Eq. (2.26) as the sum of the outputs of two parallel blocks (Fig. 18). Loop closure, of course, is implicit in the transfer functions of Fig. 18.

If a linear servo system operates in the presence of a constant disturbance, or if a regulator operates around a set point other than zero, the dynamic behavior of the system is completely unaffected. Only the steady-state operating point is altered. To control-system engineers primarily interested in dynamic performance, a change in steady-state operating point represents

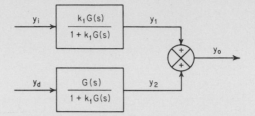

FIG. 18. COMBINED SERVO AND REGULATOR OPERATION

only a trivial scale-shifting operation. Hence they routinely take $y_d = 0$ for servo studies, $y_i = 0$ for regulator studies, and usually do not bother to treat the operating-point adjustment explicitly at all.

This casual treatment of the steady state horrifies the physiologist. To him, the value of the steady-state operating point often means the difference between life and death, and he can scarcely agree that this is a matter of indifference! In fact, his entire interest may be focused on steady-state behavior. It will be helpful, therefore, to treat explicitly the system of Fig. 16 when it operates as a regulator with a set point $y_i > 0$. The system as it stands could so function, but it would do so at a disadvantage, i.e., it would have a steady-state error for all values of y_d including $y_d = 0$. To avoid this, let us "bias" the controller by adding a reference value y_r to its output. We define this as the controller output required to zero the steady-state error when $y_d = 0$. Our new controller equation becomes:

$$y_c = k_1(y_i - y_o) + y_r, \tag{2.34}$$

where y_i is a constant set point and y_r is a constant reference value. In the present case, it is obvious that $y_r = y_i$, and the closed-loop equation in classical form becomes:

$$\left(\frac{\tau}{1 + k_1}\right)\frac{dy_o}{dt} + y_o = y_i + \left(\frac{1}{1 + k_1}\right)y_d. \tag{2.35}$$

Equation (2.35) indicates that now, when $y_d = 0$, $y_o = y_i$ in the steady state. It also indicates that there has been no change in dynamic performance, for the time constant of (2.35) is identical with that of (2.21).

The block diagram for this regulator is shown in Fig. 19.

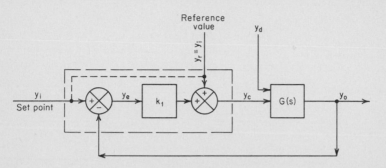

FIG. 19. REGULATOR WITH SET-POINT AND REFERENCE VALUE

In terms of physical hardware, the controller unit may have two separate knobs, one for set-point adjustment and the other for reference-value adjustment. Alternatively, by a suitable internal connection (dashed line), a single knob may serve both functions. Because we have converted the controller output, which is actually a force, F_c (Fig. 13), into equivalent displacement units y_c, in Figs. 16 and 19, $y_r = y_i$, and the distinction between set point and reference value may appear somewhat blurred. In the original notation of Fig. 13 [Eqs. (2.17) and (2.18)], the set point remains y_i but the reference value is now $F_r = Ky_i$ so that the distinction may be clearer. In either case, the difference in their functions should be clear enough: the set point is the desired output which is compared with the actual output to generate an error signal; the reference value is the value of the controller output required to zero the steady-state error when $y_d = 0$.

Summary

In its passage through a physical system, a signal is altered in two ways: in size and in timing. This is true whether we deal with open-loop systems or closed-loop feedback control systems, and whether the system is composed of mechanical, electrical, or other physical hardware. The mathematical machinery for describing such dynamic behavior involves differential equations and is often conveniently expressed in terms of transfer functions. In the next chapter, we therefore propose to examine this machinery in somewhat more detail.

CHAPTER 3

Mathematical Machinery for Linear, Lumped-Constant Physical Systems

General form of system law

THE physical systems introduced in Chapter 2 all belong to a class whose laws can be expressed in the following general mathematical form:

$$a_n \frac{d^n y}{dt^n} + a_{n-1} \frac{d^{n-1} y}{dt^{n-1}} + \ldots + a_1 \frac{dy}{dt} + y = F(t), \qquad (3.1)$$

where $F(t)$ is the system input (or forcing), y is the system output (or response), t is time, and the a's are constants. Equation (3.1) has an impressive name. It is called a linear, ordinary, non-homogeneous differential equation of nth order with constant coefficients! It is a differential equation because it contains derivatives; it is linear because none of its terms involve powers or products of the dependent variable y or its derivatives; it is ordinary because it has only one independent variable t; it is nonhomogeneous because it has one term, placed to the right of the equal sign, which differs from all the rest in not containing y or its derivatives; it is nth order because the highest derivative $d^n y / dt^n$ is an nth derivative; it has constant coefficients because the a's do not vary with time.

The most important part of this characterization is that of linearity. Although linear differential equations constitute a minute fraction of the totality of differential equations, they are the only ones for which a complete analytical theory exists and for which general analytical solutions can be obtained. It is therefore fortunate that most physical systems are sufficiently linear to permit their satisfactory description by linear differential equations.

The term ordinary also deserves comment. Equation (3.1) assumes that the properties of the system it describes are concentrated at a single point in space so that the value of y depends only upon a time coordinate and not at all upon space coordinates. Such systems are called lumped-constant systems. In contrast to these are distributed systems. Here the system properties extend over spatial coordinates which are of significant magnitude in relation to the rate of energy transmission through the system (e.g., a long electrical or pneumatic transmission line). In such systems y not only varies with time but also with space, so that the system must be described by partial differential equations which include both temporal and spatial coordinates as independent variables. We shall not consider distributed systems in this work.

Formulation of the differential equation

As we have seen in Chapter 2, the differential equation of a particular physical system can usually be obtained by applying a few relatively simple physical principles in a straightforward manner. As the number of components increases and the system grows larger, the process becomes more difficult but the general principles remain the same.

Form of the system input or forcing function

Although $F(t)$ in Eq. (3.1) may be any arbitrary function of time, certain standard input functions are used in control-system analysis and synthesis. The first of these is the unit impulse, sometimes referred to as the Dirac delta function. It is defined as the limit of a pulse of magnitude $1/a$ and duration a, as $a \to 0$ (Fig. 20a). The second function is the unit step function, also called the Heaviside unit function after the English electrical engineer who popularized its use. It is defined as a function whose value is zero for $t < 0$ and one for $t \geqq 0$ (Fig. 20b). The third function is the unit ramp function defined as a direct proportion of unit slope (Fig. 20c). These first three input functions are used in the so-called transient analysis which we shall consider in detail in Chapter 4. Note that they

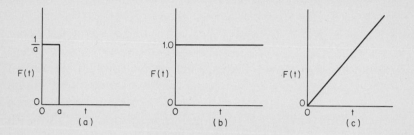

FIG. 20. (a) *UNIT IMPULSE;* (b) *UNIT STEP FUNCTION;* (c) *UNIT RAMP FUNCTION*

are interrelated, i.e., the unit ramp is the integral of the unit step, and the latter in turn is the integral of the unit impulse. The fourth function is the unit sinusoid, $F(t) = \sin \omega t$ (Fig. 21), which is used in the frequency analysis to be considered in Chapter 5. Although these functions have all been given unit

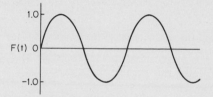

FIG. 21. *UNIT SINUSOID*

magnitude for convenience, they can be assigned any arbitrary magnitude desired.

Solution of the differential equation

As noted in Chapter 2, a solution of Eq. (3.1) will consist of a functional relationship $y(t)$, free of derivatives, between the dependent variable y and the independent variable t. The system input $F(t)$ and the system output $y(t)$ will thus be implicitly related through the common independent variable t. In general, there will be many different functions $y(t)$ which will satisfy Eq. (3.1) and hence many different solutions. The general solution is that from which all others, called particular solutions, can be obtained by specialization of arbitrary constants. Let us now examine two methods of obtaining such solutions.

Classical methods. Solution of (3.1) by the classical method involves, first, the recognition that the general solution $y(t)$ will consist of the sum of two components, $y_c(t)$ and $y_p(t)$:

$$y(t) = y_c(t) + y_p(t). \tag{3.2}$$

What these two components are called depends upon whether the caller is a mathematician or a systems engineer. The former names them, respectively, complementary function and particular integral, the latter, transient* (or force-free) response and forced response.

Now $y_c(t)$ is the general solution of the homogeneous (or force-free) equation obtained by replacing $F(t)$ in (3.1) by zero:

$$a_n \frac{d^n y}{dt^n} + a_{n-1} \frac{d^{n-1} y}{dt^{n-1}} + \ldots + a_1 \frac{dy}{dt} + y = 0. \tag{3.3}$$

The general solution of (3.3) consists of the sum of n exponential terms:

$$y_c(t) = C_1 e^{r_1 t} + C_2 e^{r_2 t} + \ldots + C_n e^{r_n t}, \tag{3.4}$$

where $C_1, C_2, \ldots C_n$ are arbitrary constants and $r_1, r_2, \ldots r_n$ are the n unequal roots of the algebraic characteristic equation obtained from (3.3) by substituting ascending powers of r for y and its derivatives (i.e., r^0 for y, r^1 for dy/dt, r^2 for d^2y/dt^2, etc.):

$$a_n r^n + a_{n-1} r^{n-1} + \ldots + a_1 r + 1 = 0. \tag{3.5}$$

If some of the roots of (3.5) are equal (say k of the n roots equal r_1), then for every such set of equal roots Eq. (3.4) will contain a modified term of the following type:

$$(C_1 + C_2 t + \ldots + C_k t^{k-1}) e^{r_1 t}. \tag{3.6}$$

The second component of the general solution of (3.1), $y_P(t)$, is

---

* We shall see later that this term applies rigorously only to the complementary function of stable systems.

any particular solution of (3.1). There are perfectly general methods of obtaining $y_P(t)$ which work for any form of $F(t)$ (e.g., variation of parameters). However, in most cases of practical interest, $F(t)$ will consist of one or more terms of the following type: constant, positive integral powers (t^n), real exponentials (e^{kt}), or sinusoids. In such cases $y_P(t)$ is most easily obtained by a process of educated guessing known formally as the method of undetermined coefficients. What we do is assume that $y_P(t)$ will have the same form as $F(t)$ as summarized in Table 2. We then sub-

TABLE 2

$F(t)$	Form of $y_P(t)$[a]
Constant, b	B
Power, bt^n	$B_0t^n + B_1t^{n-1} + \ldots + B_{n-1}t + B_n$
Real exponential, be^{kt}	Be^{kt}
Sine, $b \sin \omega_f t$	$B_0 \cos \omega_f t + B_1 \sin \omega_f t$
Cosine, $b \cos \omega_f t$	$B_0 \cos \omega_f t + B_1 \sin \omega_f t$

[a] If a term of $y_P(t)$ also appears in the complementary function, then we must use $ty_P(t)$ instead of $y_P(t)$.

stitute this form along with its derivatives into our original Eq. (3.1) and evaluate the B constants. Having done so, we can then write the general solution of (3.1) as:

$$y(t) = C_1e^{r_1t} + C_2e^{r_2t} + \ldots + C_ne^{r_nt} + y_P(t), \qquad (3.7)$$

where we have assumed that the n roots of the characteristic equation (3.5) are all unequal and that the form and constants of $y_P(t)$ have been determined as described above.

The final step in the solution of a particular problem is the evaluation of the C constants in (3.7). Since there are n of these, we shall have to provide n simultaneous equations to evaluate them. We do this by specifying n initial conditions for $y(t)$ and its derivatives, i.e., we say that when $t = 0$, $y = y_0$, $dy/dt = (dy/dt)_0$, $\ldots d^{n-1}y/dt^{n-1} = (d^{n-1}y/dt^{n-1})_0$. Equation (3.7) provides one of our required equations, and if we differentiate (3.7) $n - 1$ times, we will provide the rest. This completes the solution by the classical method. We note that it requires the solution of an nth degree algebraic characteristic equation and that the initial conditions are involved in a rather complex fashion.

The Laplace transform method.* In modern engineering practice, linear differential equations are most often solved by one of the so-called operational methods (Fourier, Laplace, Heaviside). Of these, the Laplace transform method has become an integral part of the language and technique of control-system engineering. Although this method offers no particular advantage over the classical method in solving a single nth-order differential equation with constant coefficients (e.g., both require the solution of an nth-degree algebraic equation), it does have considerable advantage in dealing with systems of equations in several dependent variables. Moreover, the fact that this method isolates the effect of the initial conditions in a single term provides the basis for the transfer-function approach to system analysis and synthesis.

In essence, the solution of a linear, ordinary differential equation by the Laplace transform method consists of transforming the differential equation in the real variable t into an algebraic equation in the complex variable s (direct Laplace transform, $\mathcal{L}$), manipulating the latter algebraically to obtain a solution in terms of s, and finally transforming this solution back into the real time domain (inverse Laplace transform, $\mathcal{L}^{-1}$). In practice, both direct and inverse transformations are accomplished by using tables of transforms for operations and functions. We shall not attempt to present the theory of the method in any detail here but will simply show how it is applied and point out some of its useful features.

Starting again with Eq. (3.1), our first step is to take the Laplace transform of both sides, and this operation is symbolically indicated as follows:

$$\mathcal{L}\left(a_n \frac{d^n y}{dt^n} + a_{n-1} \frac{d^{n-1}y}{dt^{n-1}} + \ldots + a_1 \frac{dy}{dt} + y \right) = \mathcal{L}F(t). \quad (3.8)$$

The linear properties of the transform permit (3.8) to be written alternatively as:

$$a_n\mathcal{L}\frac{d^n y}{dt^n} + a_{n-1}\mathcal{L}\frac{d^{n-1}y}{dt^{n-1}} + \ldots + a_1\mathcal{L}\frac{dy}{dt} + \mathcal{L}y = \mathcal{L}F(t). \quad (3.9)$$

* M. F. Gardner and J. L. Barnes, *Transients in Linear Systems* (New York, John Wiley and Sons, Inc., 1942); G. S. Brown and D. P. Campbell, *Principles of Servomechanisms* (New York, John Wiley and Sons, Inc., 1948).

Now, from a transform table for operations, we find that:

$$\mathcal{L}\frac{d^n y}{dt^n} = s^n \mathcal{L}y - \sum_{k=1}^{n} s^{n-k} \frac{d^{k-1}y_0}{dt^{k-1}}, \qquad (3.10)$$

where y_0 and its derivatives are the initial conditions at $t = 0$. For example, if $n = 3$, then:

$$\mathcal{L}\frac{d^3 y}{dt^3} = s^3 \mathcal{L}y - \left(s^2 y_0 + s\frac{dy_0}{dt} + \frac{d^2 y_0}{dt^2}\right). \qquad (3.11)$$

Substituting (3.10) into (3.9), collecting terms in $\mathcal{L}y$, rearranging, and using a simplified notation $[\mathcal{L}y \equiv y(s); \mathcal{L}F(t) \equiv F(s);$ all initial condition terms lumped], we get:

$$(a_n s^n + a_{n-1}s^{n-1} + \ldots + a_1 s + 1)y(s)$$
$$= F(s) + \Sigma \text{ (all ic transform terms).} \qquad (3.12)$$

Solving for $y(s)$:

$$y(s) = \left[\frac{1}{a_n s^n + a_{n-1}s^{n-1} + \ldots + a_1 s + 1)}\right]$$
$$[F(s) + \Sigma \text{ (all ic transform terms)}]. \qquad (3.13)$$

Finally, if we take the inverse transform of both sides of (3.13), our problem will be completely solved, initial conditions and all. Before attempting this, however, let us examine Eq. (3.13) more closely because it reveals why the Laplace transform method has become so popular among control-system engineers.

According to Eq. (3.13), the output transform $y(s)$ is equal to the product of two other transform terms. The first of these is what the engineers call the system transfer function (or simply the transfer function), and the second is the excitation function. The transfer function contains all of the essential information about the system itself; the excitation function contains all of the essential information about the excitation applied to the system, and includes both an external forcing (or driving) transform and an initial condition transform (or initial excitation function). If the initial conditions are all zero, then the product of transfer function and forcing transform yields the normal response transform, i.e., the response to $F(s)$ starting from rest. It should now be apparent that if in Eqs. (2.8) and (2.15) of Chapter 2, we specify that y is really $y(s)$, that F is really $F(s)$, and that the initial conditions are all zero, then the notational trick used to obtain these equations really amounts to the application of the direct Laplace transform operation to the original differential equations (2.5) and (2.13). The normal response is the one usually dealt with in control-system

FIG. 22. BLOCK DIAGRAM WITH LAPLACE NOTATION

theory and it lends itself very nicely to the block-diagram representation. Thus, if $F(s)$ is the forcing transform, $G(s)$ is the transfer function, and $y(s)$ is the output transform, our block diagram would be that of Fig. 22, and $y(s) = G(s)F(s)$.

Expressed in this form, it is very easy to deal with several

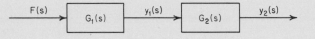

FIG. 23. BLOCKS IN CASCADE

such blocks arranged in a larger system. For example, suppose we had the series or "cascade" system shown in Fig. 23. It is easy to show that $y_2(s) = G_1(s)G_2(s)F(s)$. Again, if we had the parallel system shown in Fig. 24, it is apparent that $y_3(s) = [G_1(s) + G_2(s)]F(s)$. Finally, if we had the feedback

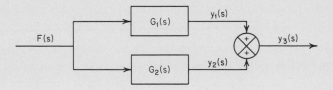

FIG. 24. BLOCKS IN PARALLEL

system of Fig. 25, it should now be apparent that the open-loop transfer function $y_o(s)/y_e(s)$ is $G_1(s)G_2(s)$, the closed-loop transfer function $y_o(s)/y_i(s)$ is $G_1(s)G_2(s)/1 + G_1(s)G_2(s)$, and the error-transfer function $y_e(s)/y_i(s)$ is $1/1 + G_1(s)G_2(s)$. These properties of the transfer function make it a very convenient tool for dealing with systems containing many "blocks" arranged in various ways. We shall say more about this later, but let us now return to the problem of applying the inverse transform operation to Eq. (3.13).

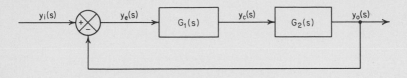

FIG. 25. FEEDBACK SYSTEM

We may indicate this operation symbolically as follows:

$$\mathcal{L}^{-1}y(s) = y(t) = \mathcal{L}^{-1}\left\{\left[\frac{1}{(a_ns^n + a_{n-1}s^{n-1} + \ldots + a_1s + 1)}\right]\right.$$

$$\left.\left[F(s) + \Sigma \text{ (all ic transform terms)}\right]\right\}. \quad (3.14)$$

We note first that the denominator of the transfer function will always be an algebraic function of s. The same will be true of the excitation function provided $F(s)$ is restricted to those forms listed in Table 1. Hence the right side of (3.14) will always be an algebraic function of s which can be manipulated like any other algebraic function. The object of such manipulation in the present case is to obtain a recognizable function of s which can be transformed back into the time domain by inspection. Although there are various ways of doing this, we shall describe only the one of most general application. It utilizes the following facts.

1. The function of s to be transformed, $y(s)$, can always be written as a rational proper fraction in lowest terms, $A(s)/Q(s)$, in which the coefficient of the highest degree term in $Q(s)$ is unity. Since $Q(s)$ is a polynomial of degree n it can be written as the product of n linear factors, $(s - s_1)(s - s_2) \ldots (s - s_n)$, where $s_1, s_2, \ldots s_n$ are the n roots of the equation $Q(s) = 0$. Some or all of these roots may be equal.

2. The function $A(s)/Q(s)$ can therefore be expanded into the sum of a number of partial fractions in the following way: If s_1 occurs once as a root of $Q(s) = 0$, it will contribute the term, $C/(s - s_1)$, where $C = (s - s_1)A(s)/Q(s)]_{s=s_1}$, or $A(s)/Q'_{s_1}(s)]_{s=s_1}$, where $Q'_{s_1}(s)$ is the product of all factors in $Q(s)$ except $(s - s_1)$.

If s_2 occurs m times as a root of $Q(s) = 0$, it will contribute the following m terms to the expansion:

$$\frac{C_1}{(s - s_2)^m} + \frac{C_2}{(s - s_2)^{m-1}} + \cdots + \frac{C_m}{(s - s_2)},$$

where

$$C_1 = \frac{(s - s_2)^m A(s)}{Q(s)}\bigg]_{s=s_2} = \frac{A(s)}{Q'_{s_2}(s)}\bigg]_{s=s_2}$$

$$C_2 = \frac{d}{ds}\left[\frac{A(s)}{Q'_{s_2}(s)}\right]_{s=s_2}$$

$$C_m = \frac{d^{m-1}}{ds^{m-1}}\left[\frac{A(s)}{Q'_{s_2}(s)}\right]_{s=s_2},$$

where $Q'_{s_2}(s)$ is the product of all factors in $Q(s)$ except $(s - s_2)^m$.

3. $\mathcal{L}^{-1}\left[\dfrac{K}{(s - s_1)}\right] = Ke^{s_1 t}$

4. $\mathcal{L}^{-1}\left[\dfrac{K}{(s - s_1)^m}\right] = \dfrac{K}{(m - 1)!}t^{m-1}e^{s_1 t}.$

Thus the inverse transform operation indicated in (3.14) resolves itself into the following steps: (1) writing $y(s)$ in standard form $A(s)/Q(s)$, (2) performing a partial fraction expansion of $A(s)/Q(s)$ as described above, and (3) transforming each partial fraction term into an exponential function of t utilizing the transform pairs in 3 and/or 4 above. This completes the solution of Eq. (3.1) by the Laplace transform method. Note that like the classical method we must solve an algebraic equation of degree n to get the roots of $Q(s) = 0$, but that unlike the classical method the initial conditions enter the solution in a simple, automatic way.

Let us note some important and close relationships between the classical and Laplace solutions of the homogeneous equation (3.3). If we take $F(s) = 0$ in (3.13) and do not specialize our initial conditions, then $\mathcal{L}^{-1}y(s)$ will be identical in form with the complementary function of the classical solution, and the exponential constants will depend upon the roots of the same characteristic equation, i.e., upon the roots of $a_n s^n + a_{n-1} s^{n-1} + \ldots + a_1 s + 1 \equiv a_n Q(s) = 0$. The only difference between the two is that in the Laplace solution the arbitrary C constants will already be expressed as general functions of the initial conditions. The denominator of the transfer function, which is identical except for a difference in notation (s instead of r) with the left side of the characteristic equation (3.5), is called the characteristic function of the system. It is then clear that the roots of the characteristic equation are the "zeros" of the characteristic function and the "infinities" or "poles" of the transfer function.

Finally, let us point out an interesting relationship between the normal response to a unit impulse, $\delta(t)$, and the force-free response. It happens that the Laplace transform of $\delta(t)$ is unity. It is therefore apparent that the right side of Eq. (3.14) will be identical under the following two conditions: (1) $F(s) = \mathcal{L}\delta(t)$, with all initial conditions zero, or (2) $F(s) = 0$, with initial conditions zero except for $(d^{n-1}y/dt^{n-1})_0 = 1/a^n$. Hence the normal response of an nth-order system to a unit impulse is identical with its force-free response from an initial condition of $1/a^n$ on the $(n-1)$th derivative. Moreover, it now becomes clear that the response to a unit impulse with arbitrary initial conditions is just the force-free response with $1/a^n$ added to the initial condition on the $(n-1)$th derivative. These considerations are related to the question of stability which we now wish to consider.

The force-free solution and stability. Engineers are vitally concerned with what is called the stability of a system. A stable system is one whose output will always return to zero in the absence of input. If a transient input (e.g., an impulse) is applied to such a system, the output will also be a transient, i.e., it will decay to zero as t approaches infinity. On the contrary, if an impulse forcing is applied to an unstable system, the output may remain at a constant value different from zero, it may oscillate continuously at a constant amplitude, or it may increase indefinitely with or without oscillation until the system is destroyed. It is therefore apparent why engineers want the systems they design to be stable.

In mathematical terms, the requirement for system stability may be stated very simply: A system will be stable if the general solution of its homogeneous (or force-free) equation, (3.3), is a transient, i.e., if its complementary function really is a transient response. We have already noted that this force-free solution can be interpreted as the response to a transient input, i.e., to an impulse. Now the complementary function will be a transient *provided all of the roots of the characteristic equation (or poles of the transfer function) have negative real parts.* Let us see how this comes about.

We have already described the forms of the complementary function, (3.4) and (3.6), and their relationship to the roots of the characteristic equation (3.5). In general, these roots or zeros may be real numbers, pure imaginary numbers, or complex numbers, the latter including the other two as special cases. Thus we recall that the complex number s is defined by:

$$s = \alpha + j\omega, \tag{3.15}$$

where the real number α is its real part, the real number ω is its imaginary part, and $j \equiv (-1)^{1/2}$. If $\omega = 0$, $s = \alpha$, a real number; if $\omega = 0$, $s = j\omega$, a pure imaginary number; if $\alpha = \omega = 0$, $s = 0$. Let us now examine the terms which appear in the complementary function for these various types of roots.

For each distinct real root, say α_1, the complementary function will contain a term which will be a decaying exponential, $Ce^{-\alpha_1 t}$, if the root is negative, an increasing exponential, $Ce^{\alpha_1 t}$, if the root is positive, and a constant, $Ce^{0t} = C$, if the root is zero. Only the first of these is a transient.

Pure imaginary roots always occur in conjugate pairs, i.e., $\pm j\omega$. For each such distinct pair, say $\pm j\omega_1$, the complementary function will contain the terms $C_1 e^{j\omega_1 t} + C_2 e^{-j\omega_1 t}$. Now it can be shown by methods we need not consider here that such a conjugate pair of imaginary exponentials corresponds to a harmonic oscillation of constant amplitude and angular frequency, i.e., $A \cos \omega_1 t$ or $A \sin \omega_1 t$. Hence pure imaginary roots introduce constant-amplitude oscillatory terms into the force-free solution. These are not transients.

Complex roots also always occur in conjugate pairs of the form $\alpha \pm j\omega$. For each such distinct pair, say $\alpha_1 \pm j\omega_1$, the complementary function will contain the terms $C_1 e^{(\alpha_1 + j\omega_1)t} + C_2 e^{(\alpha_1 - j\omega_1)t}$. If we factor out the term, $e^{\alpha_1 t}$, to get $e^{\alpha_1 t}(C_1 e^{j\omega_1 t} + C_2 e^{-j\omega_1 t})$, the nature of the solution becomes evident. Thus we already know that the bracket term represents a constant-amplitude harmonic oscillation and that $e^{\alpha_1 t}$ is a decaying exponential if α_1 is negative, and an increasing exponential if α_1 is positive. Taking the product, it is apparent that each distinct pair of conjugate complex roots corresponds to a harmonic oscillation of angular frequency ω_1 whose amplitude decays exponentially if α_1 is negative, but grows exponentially if α_1 is positive. The former represents a transient, the latter a destructive oscillation.

Finally, we need to consider the effect of repeated roots (i.e., equal roots or roots of "multiplicity > 1"). The key lies in the modified exponential term given in (3.6). It tells us that for each repeated root, say r_1, the corresponding exponential term in the complementary function must be multiplied by the factor $(C_1 + C_2 t + \ldots + C_k t^{k-1})$ where k is the multiplicity of the root. From the stability standpoint, the only important effect of this is that repeated roots of zero introduce the increasing terms $C_2 t + \ldots + C_k t^{k-1}$, and that repeated pure imaginary roots introduce growing oscillations. Repeated negative real roots, or repeated complex roots with negative real parts are still transients, for it can be shown that $\lim_{t \to \infty} t^k e^{-\alpha t} = 0$.

Table 3 provides a summary of the terms which appear in the complementary function for the various types of roots of the characteristic equation. In the middle column of the table, we have represented the roots graphically as points in the complex plane or s-plane. This is a rectangular coordinate system in which the horizontal axis is the axis of reals and the vertical axis is the axis of imaginaries. In terms of the s-plane, we can say that *the roots of the characteristic equation of a stable system all lie to the left of the imaginary axis, i.e., in the left half-plane.* We shall

TABLE 3. RELATION OF THE POLES OF THE TRANSFER FUNCTION TO THE TIME RESPONSE

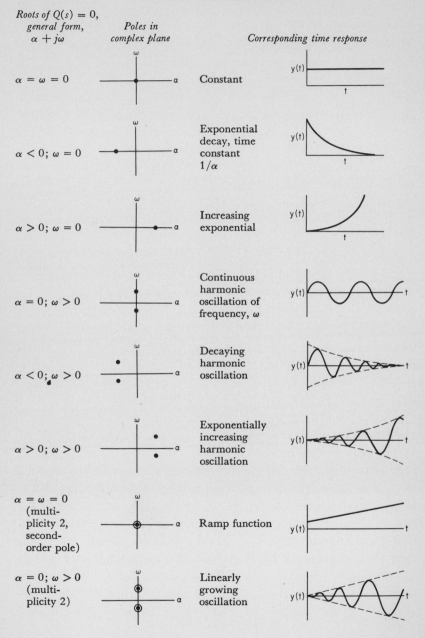

Roots of $Q(s) = 0$, general form, $\alpha + j\omega$	*Poles in complex plane*	*Corresponding time response*
$\alpha = \omega = 0$		Constant
$\alpha < 0; \omega = 0$		Exponential decay, time constant $1/\alpha$
$\alpha > 0; \omega = 0$		Increasing exponential
$\alpha = 0; \omega > 0$		Continuous harmonic oscillation of frequency, ω
$\alpha < 0; \omega > 0$		Decaying harmonic oscillation
$\alpha > 0; \omega > 0$		Exponentially increasing harmonic oscillation
$\alpha = \omega = 0$ (multiplicity 2, second-order pole)		Ramp function
$\alpha = 0; \omega > 0$ (multiplicity 2)		Linearly growing oscillation

encounter this s-plane representation frequently in later chapters.

The normal response and the unit impulse. The normal response of a linear system to an arbitrary input has an interesting interpretation in terms of its response to a unit impulse. In Laplace transform notation, the normal response to an arbitrary input $F(t)$ is:

$$y(s) = G(s)F(s) \tag{3.16}$$

and $$y(t) = \mathcal{L}^{-1}[G(s)F(s)]. \tag{3.17}$$

In particular, if $F(t)$ is the unit impulse, then $F(s) = 1$, and

$$y(s) = G(s) \tag{3.18}$$

$$y(t) = \mathcal{L}^{-1}G(s). \tag{3.19}$$

Thus the inverse Laplace transform of the transfer function $G(s)$ is the response to unit impulse $G(t)$. Now it happens that the inverse transform of (3.17) is the following real convolutional integral:

$$y(t) = \int_0^t G(\tau)F(t-\tau)d\tau. \tag{3.20}$$

Equation (3.20) says that to evaluate the output y at any instant t, the input value at all previous times $(t-\tau)$, from $\tau = 0$ to $\tau = t$, is weighted by the value $G(\tau)$ and these weighted values are summed. However, from (3.19), we note that $G(\tau)$ is the value that a unit impulse response initiated at $(t-\tau)$ would have at t. Hence $G(t)$, the normal response to unit impulse, is called the system weighting or memory function and the response to any arbitrary $F(t)$ can be built up in terms of it.

Thus we can imagine any arbitrary input to consist of a sequence of impulses of appropriate magnitudes. Each one initiates its own impulse response $KG(t-\tau)$, where K is the magnitude of the impulse, and each of these responses can be regarded as continuing to completion as if nothing else were happening. Now the actual value of $y(t)$ at any instant will be the sum of the values of all of these previously initiated impulse responses at that particular instant, and this is what Eq. (3.20) says. It is thus an expression of the superposition principle applicable to linear systems. In particular, if $F(t)$ is the unit step function, then $F(t-\tau) = 1$ in Eq. (3.20) so that

$$y(t) = \int_0^t G(\tau)d\tau$$

and the response is simply the integral of the weighting function. This relationship finds a useful biological application in indicator dilution studies on the cardiovascular system, for it reveals that a dye dilution curve resulting from a continuous infusion (step-function input) is simply the integral of that from a single injection (impulse input).

Solutions by computers. Although the analytical methods described above can be used to solve Eq. (3.1) for any value of n, they rapidly become impractical as n increases. Thus both methods require the solution of an algebraic characteristic equation of nth degree. Although general methods exist for solving third- and fourth-degree algebraic equations, they are very laborious. No general methods at all exist for solving algebraic equations of fifth degree or higher. It is in the solution of these higher-order, linear differential equations, systems of such equations in several dependent variables, and nonlinear equations for which no general analytical methods of solution exist that modern computers have great value.

Either a digital or an operational analog computer can be used to solve Eq. (3.1). The former employs one of the numerical or iterative methods of solution in which the problem of integration is converted into one of arithmetic. The same methods could be used without the computer but they would be prohibitively slow and laborious because of the great number of arithmetic calculations required. The advantage of the digital computer is that it can perform these calculations extremely rapidly (e.g., the IBM 7090 can perform 227,270 additions of seven-digit numbers *per second!*).

The analog computer uses components which perform the operations of integration, addition, subtraction, multiplication, and division on continuous variables represented by voltages. To solve Eq. (3.1) on an operational analog computer, we would first solve the equation for the highest derivative to obtain:

$$\frac{d^n y}{dt^n} = \frac{1}{a_n}\left[F(t) - \left(a_{n-1}\frac{d^{n-1}y}{dt^{n-1}} + \ldots + a_1\frac{dy}{dt} + y \right)\right]. \quad (3.21)$$

Now if we introduce a voltage equal* to $d^n y/dt^n$ as input to an

---

* For simplicity, both scaling and sign change are neglected.

integrator, the integrator output will be $d^{n-1}y/dt^{n-1}$. In symbols, we have Fig. 26 where the "barred" triangle represents an integrator. The initial condition for $d^{n-1}y/dt^{n-1}$ is introduced through the potentiometer labeled ic. Our course is now clear. We repeat this process until we have generated all of the deriva-

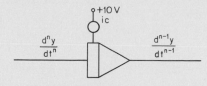

FIG. 26. INTEGRATING AMPLIFIER

tives of y as well as y itself. We multiply each by its appropriate constant (by means of potentiometers), add them, subtract the sum from $F(t)$ (which we obtain from a "function generator"), and finally multiply the result by $1/a_n$. We thus end up with $d^n y/dt^n$ which is the input to our first integrator. The simplified

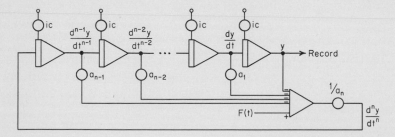

FIG. 27. ANALOG COMPUTER CIRCUIT FOR SOLUTION OF EQ. (3.21)

computer diagram is shown in Fig. 27 in which the "unbarred" triangle represents a summer.

It is apparent that these methods offer great advantages in dealing with higher-order equations or systems of equations. For nonlinear equations they may provide the only method of solution.

Summary

The laws of the linear, lumped-constant physical systems with which we shall deal are expressed in the form of linear, ordinary differential equations with constant coefficients. Such equations can be solved analytically by either the classical or the Laplace transform method. The latter has the advantage of isolating the initial conditions in an "automatic" fashion and providing a convenient transfer-function method of describing linear systems. Both methods require solution of an algebraic equation of degree equal to the order of the differential equation. The roots of this characteristic equation determine the nature of the force-free solution and, in so doing, define system stability. For higher-order systems solution of the algebraic characteristic equation becomes very laborious, and the use of digital or analog computers becomes increasingly important. In Chapter 4, we shall apply some of this mathematical machinery to a more detailed study of the physical systems introduced in Chapter 2.

CHAPTER 4

Transient Analysis of Physical Systems

IN Chapter 2 we formulated the differential equations describing particular first- and second-order physical systems and looked at some of their solutions in a preliminary way. In Chapter 3, we presented the details of the mathematical machinery required to solve these equations. With this machinery at hand, we now wish to reexamine the systems of Chapter 2 more closely. Two standard input functions are generally used to characterize the behavior of these systems. The first is the nonperiodic step function encountered in Chapter 2. A study employing this input function is called a transient analysis because the transient portion of the response is of major interest. The forced response is always a constant, and although it provides a value for steady-state gain and reveals the presence or absence of steady-state error, it can tell nothing about the dynamic properties of the system. The second standard input function is the periodic sinusoidal function. A study employing it is called a frequency analysis. It is also a steady-state analysis because here only the forced response is considered. The present chapter will consider transient analysis and Chapter 5 will explore frequency analysis.

Transient analysis of a first-order system

If for simplicity we use the symbol $\dot{y}$ for dy/dt, we can write the equation of our first-order system as follows, where F is a constant:

$$\tau\dot{y} + y = \frac{F}{K}. \tag{4.1}$$

The solution of (4.1) by the classical method is:

$$y = Ce^{rt} + y_P, \qquad (4.2)$$

where Ce^{rt} is the complementary function (or transient response) and y_P is the particular integral (or forced response). To evaluate r, we solve the algebraic characteristic equation:

$$\tau r + 1 = 0 \qquad (4.3)$$

from which $r = -1/\tau$. Since τ is positive, we know at once that the force-free solution will contain a decaying exponential with time constant τ. To evaluate y_P, we look at Table 2 and since our forcing term F/K is a constant, we take $y_P = B$, a constant. Hence $\dot{B} = 0$ and, substituting in (4.1), we have:

$$\tau(0) + B = F/K \qquad (4.4)$$

from which $y_P = F/K$. Thus our general solution of (4.1) is:

$$y = Ce^{-t/\tau} + F/K \qquad (4.5)$$

or

$$y = Ce^{-t/\tau} + y_{ss}, \qquad (4.6)$$

where $y_{ss} \equiv F/K$ is the steady-state value of y. The final step is to obtain a particular solution by evaluating the arbitrary constant C. We do this by specifying an initial condition on y, i.e., we say that when $t = 0$, $y = y_0$. Since, for $t = 0$, $e^{-t/\tau} = 1$, Eq. (4.6) becomes:

$$y_0 = C + y_{ss} \qquad (4.7)$$

from which $C = y_0 - y_{ss}$. Hence our particular solution is:

$$y = (y_0 - y_{ss})e^{-t/\tau} + y_{ss}. \qquad (4.8)$$

Note that Eq. (4.8) implies a more general concept of step-function forcing than given by our previous definitions. The essential feature of a step function is that it assumes one constant value for $t < 0$ and a different constant value for $t \geqq 0$. The former need not be restricted to zero but can be any positive or negative constant as well.

In Fig. 28, we have plotted Eq. (4.8) for $y_0 < y_{ss}$, and in Fig. 29 for $y_0 > y_{ss}$. We can include *all* first-order responses to a step function in a *single nondimensional* curve by rearranging Eq. (4.8) in the following way:

$$\frac{y - y_{ss}}{y_0 - y_{ss}} = e^{-t/\tau}. \tag{4.9}$$

Since both numerator and denominator of (4.9) have the

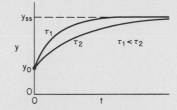

FIG. 28. FIRST-ORDER RESPONSE TO STEP FUNCTION, $y_{ss} > y_0 > 0$

same dimensions (linear displacement in this case), the ratio is dimensionless. It is the ratio of the change yet to be made to the total change. Since both t and τ have dimensions of time, their ratio is also dimensionless. Hence in Fig. 30, in which $(y - y_{ss})/(y_0 - y_{ss})$ is plotted against t/τ, both axes are dimen-

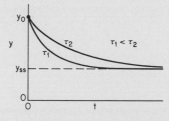

FIG. 29. FIRST-ORDER RESPONSE TO STEP FUNCTION, $y_{ss} < y_0 > 0$

sionless ratios and all first-order responses to a step function are included in this single curve. From this single dimensionless curve, we can easily construct dimensional response curves for any particular values of τ, y_0, and y_{ss}.

If we take the natural logarithms of both sides of Eq. (4.9), we get:

$$\ln\left(\frac{y - y_{ss}}{y_0 - y_{ss}}\right) = -\frac{1}{\tau} t. \tag{4.10}$$

This means that a plot of $\ln\left[(y - y_{ss})/(y_0 - y_{ss})\right]$ against t will yield a straight line whose slope is $1/\tau$. This provides a method for solving the inverse problem, i.e., for evaluating τ from the observed response to a step function.

To solve (4.1) by the Laplace transform method, we first

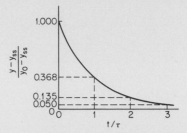

FIG. 30. NONDIMENSIONALIZED FIRST-ORDER RESPONSE TO
STEP FUNCTION

apply the direct transform to both sides. To do so we recall from Eq. (3.10) that $\mathcal{L}(\tau\dot{y}) = \tau s y(s) - \tau y_0$ and that $\mathcal{L}y(t) = y(s)$. Finally, we note from a transform table that $\mathcal{L}A$, where A is any constant, is A/s. Since the right side of (4.1) is the constant $F/K \equiv y_{ss}$, it is apparent that $\mathcal{L}F/K = (F/K)/s = y_{ss}/s$. Hence the direct transform of (4.1) is:

$$\tau s y(s) - \tau y_0 + y(s) = \frac{y_{ss}}{s} \tag{4.11}$$

and collecting terms in $y(s)$ yields:

$$(\tau s + 1)y(s) = \frac{y_{ss}}{s} + \tau y_0. \tag{4.12}$$

Dividing through by $(\tau s + 1)$ and combining the terms on the right over a common denominator we then obtain:

$$y(s) = \frac{y_{ss} + s\tau y_0}{s(\tau s + 1)}. \tag{4.13}$$

Preliminary to partial fraction expansion of (4.13), we divide numerator and denominator by τ to get:

$$y(s) = \frac{y_{ss}/\tau + s y_0}{s(s + 1/\tau)}. \tag{4.14}$$

The roots of $s(s + 1/\tau) = 0$ are obviously zero and $-1/\tau$, and partial fraction expansion of (4.14) yields:

$$y(s) = \frac{y_{ss}}{(s - 0)} + \frac{y_0 - y_{ss}}{[s - (-1/\tau)]}. \qquad (4.15)$$

Finally, applying the inverse transform, we obtain Eq. (4.8) as before:

$$y(t) = (y_0 - y_{ss})e^{-t/\tau} + y_{ss}. \qquad (4.8)$$

Note that in the Laplace procedure the initial conditions are automatically included in the transform process. As noted above, the poles of (4.14) are zero and $-1/\tau$. The former arises from the forcing function and contributes the constant term y_{ss} to

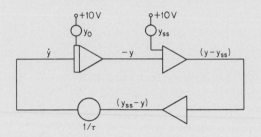

FIG. 31. ANALOG COMPUTER CIRCUIT FOR SOLUTION OF EQ. (4.1)

the time solution. The latter arises from the system transfer function $[1/(\tau s + 1)]$ and contributes a decaying exponential (the transient response) to the time solution. In the transfer-function notation of Chapters 2 and 3, we would write Eq. (4.13) as:

$$y(s) = \left[\frac{1}{\tau s + 1}\right]\left[\frac{y_{ss}}{s} + \tau y_0\right]. \qquad (4.16)$$

Finally, to solve (4.1) using an operational analog computer, we would write:

$$\dot{y} = \frac{1}{\tau}(y_{ss} - y). \qquad (4.17)$$

The computer circuit is shown in Fig. 31. It includes three operational amplifiers (one serving as an integrator, one as a summer, and one as a sign changer), and three potentiometers which serve to set the values of y_0, y_{ss}, and τ, respectively. In the circuit of Fig. 31, we have included the fact (previously neglected for simplicity in Fig. 27) that an operational amplifier always changes the sign of the input voltages.

Transient analysis of a second-order system

If we let $\ddot{y} \equiv d^2y/dt^2$, $\dot{y} \equiv dy/dt$, and $y_{ss} = F/K$ (a constant), we can write the equation of our second-order system as follows:

$$\frac{1}{\omega_n{}^2}\ddot{y} + \frac{2\zeta}{\omega_n}\dot{y} + y = y_{ss}. \tag{4.18}$$

Solution of (4.18) by either the classical or Laplace methods will require determination of the roots of the following characteristic equation:

$$\frac{1}{\omega_n{}^2}r^2 + \frac{2\zeta}{\omega_n}r + 1 = 0. \tag{4.19}$$

Applying the familiar quadratic formula to (4.19), these roots are:

$$r_1 = -\zeta\omega_n + \omega_n(\zeta^2 - 1)^{1/2} \tag{4.20}$$
$$r_2 = -\zeta\omega_n - \omega_n(\zeta^2 - 1)^{1/2}. \tag{4.21}$$

Recalling our discussion of stability in Chapter 3, it is at once evident that the nature of the solution will be critically dependent upon the value of the damping ratio ζ, for this will determine whether the roots are conjugate imaginary ($\zeta = 0$), conjugate complex ($0 < \zeta < 1$), real and equal ($\zeta = 1$), or real and distinct ($\zeta > 1$).

If the roots of (4.19) are distinct, the general solution of (4.18) by the classical method is:

$$y = C_1 e^{r_1 t} + C_2 e^{r_2 t} + y_{ss}. \tag{4.22}$$

If the roots of (4.19) are equal (as they will be for $\zeta = 1$), the general solution of (4.18) is:

$$y = (C_3 + C_4 t)e^{r t} + y_{ss}. \tag{4.23}$$

To evaluate C_1 and C_2 in (4.22), we assign the two initial conditions, $y = y_0$ and $\dot{y} = 0$ at $t = 0$. Then (4.22) becomes:

$$y_0 = C_1 + C_2 + y_{ss}. \tag{4.24}$$

Differentiation of (4.22) yields:

$$\dot{y} = C_1 r_1 e^{r_1 t} + C_2 r_2 e^{r_2 t} \qquad (4.25)$$

which, at $t = 0$, becomes:

$$0 = C_1 r_1 + C_2 r_2. \qquad (4.26)$$

Simultaneous solution of (4.24) and (4.26) then yields the values of C_1 and C_2:

$$C_1 = (y_0 - y_{ss})\left(\frac{r_2}{r_2 - r_1}\right) \qquad (4.27)$$

$$C_2 = (y_0 - y_{ss})\left(\frac{r_1}{r_1 - r_2}\right). \qquad (4.28)$$

Substituting these values into (4.22) yields our particular solution:

$$y = (y_0 - y_{ss})\left(\frac{r_2}{r_2 - r_1}\right)e^{r_1 t} + (y_0 - y_{ss})\left(\frac{r_1}{r_1 - r_2}\right)e^{r_2 t} + y_{ss}. \quad (4.29)$$

Once more we can write (4.29) in terms of the dimensionless ratio $(y - y_{ss})/(y_0 - y_{ss})$ as follows:

$$\frac{y - y_{ss}}{y_0 - y_{ss}} = \left(\frac{r_2}{r_2 - r_1}\right)e^{r_1 t} + \left(\frac{r_1}{r_1 - r_2}\right)e^{r_2 t}. \qquad (4.30)$$

To evaluate C_3 and C_4 in (4.23), we apply the same initial conditions. At $t = 0$, (4.22) then becomes:

$$y_0 = C_3 + y_{ss} \qquad (4.31)$$

from which:

$$C_3 = y_0 - y_{ss} \qquad (4.32)$$

the derivative of (4.23) is:

$$\dot{y} = (C_3 + C_4 t)r_1 e^{r_1 t} + C_4 e^{r_1 t} \qquad (4.33)$$

which, at $t = 0$, becomes:

$$0 = C_3 r_1 + C_4 \qquad (4.34)$$

and substituting the value of C_3 from (4.32) we evaluate C_4:

$$C_4 = -(y_0 - y_{ss})r_1. \qquad (4.35)$$

Substituting (4.32) and (4.35) into (4.23) yields our particular solution:

$$y = (y_0 - y_{ss})(1 - r_1 t)e^{r_1 t} + y_{ss} \qquad (4.36)$$

which in dimensionless form is:

$$\frac{y - y_{ss}}{y_0 - y_{ss}} = (1 - r_1 t)e^{r_1 t}. \tag{4.37}$$

We are now ready to examine the responses of our second-order system to a step-function forcing for particular values of ζ using Eqs. (4.30) and (4.37). We shall consider four cases: $\zeta = 0$, $0 < \zeta < 1$, $\zeta = 1$, and $\zeta > 1$.

Case 1: $\zeta = 0$, the harmonic oscillator. If $\zeta = 0$, the roots of the characteristic equation as defined by (4.20) and (4.21) will be the conjugate imaginary pair $\pm j\omega_n$. Table 3 of Chapter 3 tells us that such roots are associated with a constant-amplitude harmonic oscillation of frequency ω_n. Let us see how this emerges from Eq. (4.30). Substituting $r_1 = j\omega_n$ and $r_2 = -j\omega_n$ into (4.30) yields:

$$\frac{y - y_{ss}}{y_0 - y_{ss}} = \frac{-j\omega_n}{-2j\omega_n} e^{j\omega_n t} + \frac{j\omega_n}{2j\omega_n} e^{-j\omega_n t} \tag{4.38}$$

which simplifies to:

$$\frac{y - y_{ss}}{y_0 - y_{ss}} = \frac{1}{2} \left(e^{j\omega_n t} + e^{-j\omega_n t} \right). \tag{4.39}$$

In order to express (4.38) in a more familiar form, we now make use of the Euler formulas which define imaginary exponentials in terms of trigonometric functions. They are:

$$e^{j\omega_n t} = \cos \omega_n t + j \sin \omega_n t \tag{4.40}$$
$$e^{-j\omega_n t} = \cos \omega_n t - j \sin \omega_n t. \tag{4.41}$$

It is evident that the term in parentheses in (4.39) can be evaluated in terms of the trigonometric functions simply by adding (4.40) and (4.41). The result is $(2 \cos \omega_n t)$, and substituting this in (4.39) we obtain our desired relationship:

$$\frac{y - y_{ss}}{y_0 - y_{ss}} = \cos \omega_n t. \tag{4.42}$$

Equation (4.42) is plotted in Fig. 32.

What does $\zeta = 0$ (or zero damping) mean in terms of our physical system? Recalling the definition of ζ given in Chapter 2 $[\zeta \equiv R/2(KM)^{1/2}]$, we see that $\zeta = 0$ implies $R = 0$, i.e., our system is frictionless. The energy imparted to the system by the force source is never degraded to heat since there is no energy sink. Instead it is continuously exchanged between the potential-energy storage (the spring) and the kinetic-energy storage (the mass). The potential energy of the spring is maximal when the kinetic energy of the mass is zero and vice versa. The forcing

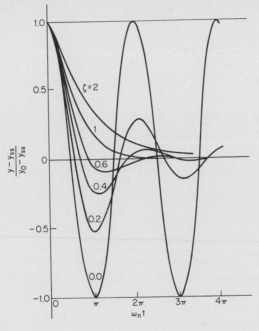

FIG. 32. SECOND-ORDER TRANSIENTS

term $y_{ss} \equiv F/K$ determines the amplitude of the harmonic oscillation.

Case 2: $0 < \zeta < 1$, the underdamped system. If $0 < \zeta < 1$, the roots of the characteristic equation will be a conjugate complex pair. It will be convenient to write them as follows:

$$r_1 = -\zeta\omega_n + j\omega \qquad (4.43)$$
$$r_2 = -\zeta\omega_n - j\omega, \qquad (4.44)$$

where $\omega = \omega_n(1 - \zeta^2)^{1/2}$. Table 3 of Chapter 3 tells us that such roots are associated with an exponentially damped harmonic oscillation. Let us see how this emerges from Eq. (4.30). Substitution of (4.43) and (4.44) into (4.30) yields:

$$\frac{y - y_{ss}}{y_0 - y_{ss}} = \left(\frac{-\zeta\omega_n - j\omega}{-2j\omega}\right) e^{(-\zeta\omega_n + j\omega)t} + \left(\frac{-\zeta\omega_n + j\omega}{2j\omega}\right) e^{(-\zeta\omega_n - j\omega)t}.$$

(4.45)

We can factor out $e^{-\zeta\omega_n t}$ to get:

$$\frac{y - y_{ss}}{y_0 - y_{ss}} = e^{-\zeta\omega_n t}\left[\left(\frac{-\zeta\omega_n - j\omega}{-2j\omega}\right) e^{j\omega t} + \left(\frac{-\zeta\omega_n + j\omega}{2j\omega}\right) e^{-j\omega t}\right].$$

(4.46)

Let us write (4.46) as:

$$\frac{y - y_{ss}}{y_0 - y_{ss}} = e^{-\zeta\omega_n t}(A_1 e^{j\omega t} + A_2 e^{-j\omega t}).$$ (4.47)

Now, by substituting the Euler relations (4.40) and (4.41) into (4.47), we can obtain the following relationship:

$$\frac{y - y_{ss}}{y_0 - y_{ss}} = e^{-\zeta\omega_n t}(B_1 \cos \omega t + B_2 \sin \omega t),$$ (4.48)

where B_1 and B_2 are defined in terms of A_1 and A_2 as follows:

$$B_1 \equiv A_1 + A_2$$ (4.49)
$$B_2 \equiv j(A_1 - A_2).$$ (4.50)

Finally, if we perform the operations indicated in (4.49) and (4.50) on the constants of (4.45), we obtain our desired relationship:

$$\frac{y - y_{ss}}{y_0 - y_{ss}} = e^{-\zeta\omega_n t}\left[\cos \omega t + \frac{\zeta}{(1 - \zeta^2)^{1/2}} \sin \omega t\right].$$ (4.51)

The interpretation of (4.51) will be more obvious if we define a new constant equal to $\{1 + [\zeta/(1 - \zeta^2)^{1/2}]^2\}^{1/2} = 1/(1 - \zeta^2)^{1/2}$ and an angle, $-\pi < \phi \leq \pi$, such that $\cos \phi = (1 - \zeta^2)^{1/2}$. In terms of these, (4.51) can be written in the following form:

$$\frac{y - y_{ss}}{y_0 - y_{ss}} = \frac{e^{-\zeta\omega_n t}}{(1 - \zeta^2)^{1/2}} \cos (\omega t - \phi).$$ (4.52)

It is clear that Eq. (4.52) describes a harmonic oscillation of angular frequency ω within a decaying exponential envelope of time constant $1/\zeta\omega_n$. Equation (4.52) is plotted in Fig. 32 for several values of ζ. Note that when $\zeta = 0$, Eqs. (4.51) and (4.52) reduce to (4.42).

Case 3: $\zeta = 1$, *the critically damped system.* When $\zeta = 1$, the roots of the characteristic equation are real and equal, i.e., $r_1 = r_2 = -\zeta\omega_n = -\omega_n$. This critical value of ζ marks the border between the oscillatory responses observed when $0 \leqq \zeta < 1$ and the nonoscillatory behavior observed when $\zeta \geqq 1$. The equation for the critically damped response is easily obtained by substituting $(-\omega_n)$ for r_1 in (4.37). The result is:

$$\frac{y - y_{ss}}{y_0 - y_{ss}} = (1 + \omega_n t)e^{-\omega_n t} \tag{4.53}$$

and this is plotted in Fig. 32.

Case 4: $\zeta > 1$, the overdamped system. If $\zeta > 1$, the roots of the characteristic equation are real, distinct, and negative, i.e., $r_1 = -\zeta\omega_n + \omega_1$ and $r_2 = -\zeta\omega_n - \omega_1$, where $\omega_1 \equiv \omega_n(\zeta^2 - 1)^{1/2}$. It turns out to be convenient to express these roots in terms of two time constants defined as follows:

$$\tau_1 \equiv -\frac{1}{r_1} \tag{4.54}$$

$$\tau_2 \equiv -\frac{1}{r_2}. \tag{4.55}$$

If we substitute (4.54) and (4.55) into (4.30), our solution for the overdamped system in terms of these time constants becomes:

$$\frac{y - y_{ss}}{y_0 - y_{ss}} = \left(\frac{\tau_1}{\tau_1 - \tau_2}\right)e^{-t/\tau_1} - \left(\frac{\tau_2}{\tau_1 - \tau_2}\right)e^{-t/\tau_2}. \tag{4.56}$$

Now let us examine this solution as ζ increases above one. From the definitions of τ_1 and τ_2, it is apparent that their ratio, τ_1/τ_2, is equal to $[\zeta + (\zeta^2 - 1)^{1/2}]/[\zeta - (\zeta^2 - 1)^{1/2}]$, and this ratio increases very rapidly with ζ. Thus for $\zeta = 1.5, 2, 3, 4$, and 5, $\tau_1/\tau_2 = 6.9, 13.8, 34.3, 60.6$, and 99. If τ_1/τ_2 is high, the second exponential term in (4.56) contributes very little to the solution since its initial amplitude is small and it dies out rapidly. Hence, as ζ increases, (4.56) approaches first-order behavior. Equation (4.56) is plotted in Fig. 32 for $\zeta = 2$.

The dimensionless curves of Fig. 32 and the equations in Table 4 conveniently summarize the solution of the direct problem for the second-order response to a step function. If $\zeta = 0$, the system oscillates continuously at its natural fre-

TABLE 4. SECOND-ORDER RESPONSES TO A STEP FUNCTION

Value of damping ratio	$\dfrac{y - y_{ss}}{y_0 - y_{ss}}$	Special definitions
$\zeta = 0$, harmonic oscillator	$\cos \omega_n t$	
$0 < \zeta < 1$, underdamped	$\left[\dfrac{e^{-\zeta \omega_n t}}{(1 - \zeta^2)^{1/2}}\right] \cos(\omega t - \phi)$	$\phi \equiv \cos^{-1}(1 - \zeta^2)^{1/2}$ $\omega \equiv \omega_n(1 - \zeta^2)^{1/2}$
$\zeta = 1$, critically damped	$(1 + \omega_n t)e^{-\omega_n t}$	
$\zeta > 1$, overdamped	$\left(\dfrac{\tau_1}{\tau_1 - \tau_2}\right)e^{-t/\tau_1} - \left(\dfrac{\tau_2}{\tau_1 - \tau_2}\right)e^{-t/\tau_2}$	$\tau_1 \equiv \dfrac{1}{[\zeta\omega_n - \omega_n(\zeta^2 - 1)^{1/2}]}$ $\tau_2 \equiv \dfrac{1}{[\zeta\omega_n + \omega_n(\zeta^2 - 1)^{1/2}]}$

quency ω_n. As ζ increases towards one, the frequency of the oscillation decreases and the amplitude is exponentially damped at an increasing rate. The oscillatory component just disappears when $\zeta = 1$ (critical damping). As ζ increases above one, the behavior of the system rapidly approaches that of a first-order system with a single time constant.

Let us briefly examine the solution of the inverse problem. The method of attack will depend upon whether the observed response is obviously oscillatory, near critical damping, or overdamped. If the response is that of a harmonic oscillator, the solution is obvious: $\zeta = 0$ and $\omega_n = 2\pi/T$, where T is the measured period of oscillation. If the response is a damped harmonic oscillation, we recall that the frequency of this oscillation is $\omega_n(1 - \zeta^2)^{1/2}$ and that the time constant of the exponential envelope is $1/\zeta\omega_n$. From the latter, it is apparent that the natural logarithm of the ratio of any positive peak amplitude to the just preceding one (called the logarithmic decrement D) is:

$$D \equiv \ln\left(\frac{y_{+p(n+1)}}{y_{+p(n)}}\right) = -\zeta\omega_n T, \qquad (4.57)$$

where T is the period of oscillation. But we know that $T = 2\pi/\omega_n(1 - \zeta^2)^{1/2}$ and, substituting in (4.57), we have:

$$D \equiv \frac{2\pi\zeta}{(1 - \zeta^2)^{1/2}} \qquad (4.58)$$

and solving for ζ:

$$\zeta \equiv \frac{D}{(D^2 + 4\pi^2)^{1/2}}. \qquad (4.59)$$

Knowing ζ and the frequency of oscillation ω it is easy to get ω_n [i.e., $\omega_n = \omega/(1 - \zeta^2)^{1/2}$]. We shall only mention the solution for the near-critical and overdamped cases. The former depends upon evaluating the time intervals required for the response to pass through the points 0.736, 0.406, and 0.199 en route to the first zero. These intervals are exactly equal for $\zeta = 1$, successively shorter for $\zeta < 1$, and successively longer for $\zeta > 1$. For an overdamped system, τ_1 and τ_2 can be evaluated from the slopes of semilogarithmic response curves. As noted previously, as ζ increases above one, the behavior rapidly approaches that of a first-order system. Knowing τ_1 and τ_2, ζ and ω_n can be calculated.

If we wish to solve (4.18) by the Laplace transform method, we first apply the direct transform to both sides to obtain:

$$\frac{1}{\omega_n^2} [s^2 y(s) - s y_0 - \dot{y}_0] + \frac{2\zeta}{\omega_n} [s y(s) - y_0] + y(s) = \mathcal{L} y_{ss}. \qquad (4.60)$$

Taking $\dot{y}_0 = 0$ as before, noting that $\mathcal{L} y_{ss} = (1/s)(y_{ss})$, and rearranging, (4.60) becomes:

$$\left(\frac{1}{\omega_n^2} s^2 + \frac{2\zeta}{\omega_n} s + 1 \right) y(s) = \frac{1}{s} (y_{ss}) + \left(\frac{1}{\omega_n^2} s + \frac{2\zeta}{\omega_n} \right) y_0 \qquad (4.61)$$

which in transfer-function form is:

$$y(s) = \left\{ \frac{1}{[(1/\omega_n^2)s^2 + (2\zeta/\omega_n)s + 1]} \right\}$$
$$\left\{ \frac{1}{s} (y_{ss}) + [(1/\omega_n^2)s + (2\zeta/\omega_n)] y_0 \right\}. \qquad (4.62)$$

The first factor on the right is the system transfer function (a second-order lag) and the second is the excitation function which includes both the external step-function input and the initial conditions. If the poles of the transfer function are distinct, partial fraction expansion of (4.62) yields:

$$y(s) = \frac{\omega_n^2 y_{ss}}{s_1 s_2 (s - 0)} + \frac{\omega_n^2 y_{ss} + (s_1^2 + 2\zeta\omega_n s_1)y_0}{s_1(s_1 - s_2)(s - s_1)}$$

$$+ \frac{\omega_n^2 y_{ss} + (s_2^2 + 2\zeta\omega_n s_2)y_0}{s_2(s_2 - s_1)(s - s_2)} \quad (4.63)$$

and applying the inverse transform:

$$y(t) = \frac{\omega_n^2 y_{ss}}{s_1 s_2} + \frac{\omega_n^2 y_{ss} + (s_1^2 + 2\zeta\omega_n s_1)y_0}{s_1(s_1 - s_2)} e^{s_1 t}$$

$$+ \frac{\omega_n^2 y_{ss} + (s_2^2 + 2\zeta\omega_n s_2)y_0}{s_2(s_2 - s_1)} e^{s_2 t}. \quad (4.64)$$

Although it is not obvious that (4.64) is identical with (4.29), a little algebraic manipulation will show that it really is. Thus, for permissible values of ζ ($\zeta = 0$ or any $+$ value except 1), $s_1 s_2 = \omega_n^2$, so that the first term is just y_{ss}. Also, one can show that $(s_1^2 + 2\zeta\omega_n s_1) = -\omega_n^2$ and that $(s_2^2 + 2\zeta\omega_n s_2) = \omega_n^2$. Then, if the second term is multiplied by (s_2/s_2) and the third by (s_1/s_1), (4.64) reduces to (4.29).

If the poles of the transfer function are equal (i.e., if $\zeta = 1$), then partial fraction expansion of (4.62) yields:

$$y(s) = \frac{\omega_n^2 y_{ss}}{s_1^2(s - 0)} + \frac{\omega_n^2 y_{ss} + (s_1^2 + 2\zeta\omega_n s_1)y_0}{s_1(s - s_1)^2} + \frac{s_1^2 y_0 - \omega_n^2 y_{ss}}{s_1^2(s - s_1)}$$

$$(4.65)$$

and applying the inverse transform:

$$y(t) = \frac{\omega_n^2 y_{ss}}{s_1^2} + \left[\frac{\omega_n^2 y_{ss} + (s_1^2 + 2\zeta\omega_n s_1)y_0}{s_1} t + \frac{s_1^2 y_0 - \omega_n^2 y_{ss}}{s_1^2}\right] e^{s_1 t}.$$

$$(4.66)$$

If we now note that $s_1 = -\zeta\omega_n = -\omega_n$, a little algebraic manipulation will reduce (4.66) to (4.36).

Finally, let us solve (4.18) with the help of an analog computer. We rearrange (4.18) to read:

$$\ddot{y} = \omega_n^2 y_{ss} - 2\zeta\omega_n \dot{y} - \omega_n^2 y \quad (4.67)$$

and set up the analog circuit of Fig. 33. The three inputs to the first integrator are $(\omega_n^2 y_{ss})$, $(-2\zeta\omega_n \dot{y})$, and $(-\omega_n^2 y)$. The integrator adds these (to get $\ddot{y}$), integrates the sum (to get $\dot{y}$), and changes the sign to finally yield $(-\dot{y})$. The rest of the circuit is self-explanatory.

Effects of feedback

In Chapter 2 we discovered that the addition of closed-loop proportional control to a first-order system did not change the form of the differential equation. However, it did modify the transient response by shortening the time constant, and a steady-state error was always present. The possibilities for modification

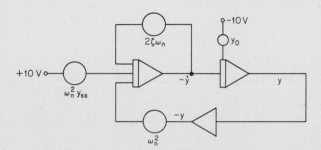

FIG. 33. ANALOG COMPUTER CIRCUIT FOR SOLUTION OF EQ. (*4.18*)

of the transient response are considerably more interesting if we add such control to a second-order system, for here there is opportunity for oscillation and instability. Let us write the equation for our controlled system as:

$$\frac{1}{\omega_n{}^2}\ddot{y} + \frac{2\zeta}{\omega_n}\dot{y} + y = y_c + y_d,\qquad(4.68)$$

where y_c and y_d are the controlling and disturbing signals, respectively. Our proportional controller equation is:

$$y_c = k_p(y_i - y)\qquad(4.69)$$

and substituting in (4.68) and rearranging, we obtain our closed-loop equation:

$$\frac{1}{\omega_n{}^2(1 + k_p)}\ddot{y} + \frac{2\zeta}{\omega_n(1 + k_p)}\dot{y} + y = \frac{k_p}{1 + k_p}y_i + \frac{1}{1 + k_p}y_d.$$
$$(4.70)$$

Comparing (4.70) with (4.68), we see that the closed-loop system is still second order but has a different effective natural frequency and damping ratio than the open-loop system. If we call these closed-loop parameters ω_{n_c} and ζ_c, we can define them in terms of the original open-loop ones, ω_n and ζ, as follows:

$$\omega_{n_c} = \omega_n(1 + k_p)^{1/2} \tag{4.71}$$

$$\zeta_c = \zeta/(1 + k_p)^{1/2}. \tag{4.72}$$

Thus the introduction of closed-loop proportional control has increased the natural frequency and decreased the damping ratio. The former is a desirable effect for, as we shall see in more detail later, it provides more effective control of high-frequency disturbances. However, the reduction of ζ may introduce undesirable oscillation. It is thus apparent that we face conflicting requirements in choosing the best value for the controller gain k_p. On the one hand, we would like to have k_p as large as possible for, as Eq. (4.70) indicates, this will minimize steady-state error for either servo or regulator operation. But, on the other hand, the larger we make k_p, the smaller will be the damping ratio ζ_c and the system may show oscillation. Thus, if we use proportional control only, we must make some compromise choice for k_p which will make the steady-state error as small as possible without introducing undesirable oscillation. However, by combining proportional control with other control modes, it is possible to improve on this compromise. Let us examine some of these combinations.

Proportional plus derivative control

If we add a term proportional to the rate of change of error to our controller equation (4.69), then the closed-loop equation will include a controller constant which can be adjusted to increase damping without simultaneously increasing steady-state error. The equation for this proportional plus derivative controller is:

$$y_c = k_p(y_i - y) + k_r(\dot{y}_i - \dot{y}). \tag{4.73}$$

Combining this with the controlled-system equation (4.68), we obtain the following closed-loop equation:

$$\frac{1}{\omega_n^2(1 + k_p)}\ddot{y} + \left[\frac{2\zeta + k_r\omega_n}{\omega_n(1 + k_p)}\right]\dot{y} + y = \frac{k_p}{1 + k_p}y_i + \frac{1}{1 + k_p}y_d$$

$$+ \frac{k_r}{1 + k_p}\dot{y}_i. \tag{4.74}$$

In terms of the controlled-system parameters ω_n and ζ, the natural frequency of the closed-loop system is again given by Eq. (4.71), but the closed-loop damping ratio ζ'_c is now:

$$\zeta'_c = \frac{2\zeta + k_r\omega_n}{2(1 + k_p)^{1/2}}. \tag{4.75}$$

The advantage gained is now apparent: we can make k_p large enough to minimize steady-state error and simultaneously maintain the desired damping ratio by increasing the derivative control gain k_r. It turns out to be true in general that derivative control will increase system damping and thereby improve stability. Note, however, that steady-state error still cannot be eliminated completely, for this would require an infinite value of the proportional gain constant k_p. Let us next examine a control mode combination which will get around this difficulty and thus allow the complete elimination of steady-state error.

Proportional plus integral control

The control mode which eliminates steady-state error is known as integral control. In this type, the rate of change of the controlling signal rather than the signal itself is made proportional to the error, i.e.,

$$\dot{y}_c = k_i(y_i - y). \tag{4.76}$$

It is apparent from Eq. (4.76) that the controller output will now continue to change in the direction required to correct the error as long as any error at all exists. The origin of the term integral control will be more obvious if we integrate (4.76) to obtain:

$$y_c = k_i \int (y_i - y)\, dt. \tag{4.77}$$

From Eq. (4.77) it is apparent that integral control action can alternatively be described by saying that the controller output is proportional to the time integral of the error. In more picturesque language, we can say that the integral controller "memorizes" the time history of the error, and that this memory, persisting after the error has been reduced to zero, is sufficient to maintain the corrective action required to keep the error at zero. This is quite different from a proportional controller which requires the presence of error to produce any corrective action at all, and from a derivative controller which produces corrective action only when the error is changing.

Although integral control (unlike derivative control) can be used alone, a faster response within acceptable stability limits can usually be obtained if it is properly combined with proportional control. The controller equation for this combined control mode is:

$$y_c = k_p(y_i - y) + k_i \int (y_i - y) \, dt. \qquad (4.78)$$

For reasons which will soon become apparent, it will be most convenient to explore the nature of this control mode by having it operate upon the first-order controlled system described in Chapter 2 and defined by the following equation:

$$\tau \dot{y} + y = y_c + y_d. \qquad (4.79)$$

Substituting (4.78) into (4.79) yields a new type of equation which, for obvious reasons, is called an integrodifferential equation:

$$\tau \dot{y} + y = k_p(y_i - y) + k_i \int (y_i - y) \, dt + y_d. \qquad (4.80)$$

To eliminate the integral in (4.80), we differentiate with respect to time to get:

$$\tau \ddot{y} + \dot{y} = k_p(\dot{y}_i - \dot{y}) + k_i(y_i - y) + \dot{y}_d \qquad (4.81)$$

and, rearranging terms, we obtain the following closed-loop equation:

$$\frac{\tau}{k_i} \ddot{y} + \frac{1 + k_p}{k_i} \dot{y} + y = y_i + \frac{k_p}{k_i} \dot{y}_i + \frac{1}{k_i} \dot{y}_d. \qquad (4.82)$$

Let us note two striking things about Eq. (4.82). First, the system which it describes will show no steady-state error for either servo or regulator operation. Thus, if y_i and y_d are step functions, $\dot{y}_i = \dot{y}_d = 0$ and $y = y_i$ in the steady state. Second, we note that the introduction of integral control has increased the order of the differential equation by one. Thus the controlled-system equation, (4.79), as well as the closed-loop equation with proportional control only [Eq. (2.21) of Chapter 2], are first order, but Eq. (4.82) is second order. Both these features of (4.82) are true in general, i.e., integral control eliminates steady-state error and increases the order of the system equation by one. It was to avoid having to deal with a third-order closed-loop equation at this point that we chose to have Eq. (4.78) operate on a first-order rather than a second-order controlled system.

Finally, another general property of integral control emerges if we recall that a first-order system may be regarded as an infinitely damped second-order system. From this point of view, we can say that the introduction of integral control has reduced the system damping. A natural frequency and damping ratio for Eq. (4.82) can be defined as follows:

$$\omega_n = \left(\frac{k_i}{\tau}\right)^{1/2} \tag{4.83}$$

$$\zeta = \frac{(1 + k_p)}{2(k_i\tau)^{1/2}}. \tag{4.84}$$

It is apparent from these definitions that as the integral controller gain k_i increases, the damping ratio falls and the natural frequency increases. In our particular example, increasing proportional controller gain k_p will increase damping for all values of k_i and no advantage is to be gained by adding derivative control. In higher-order systems, this is not generally true and derivative control is often added in order to increase system damping (proportional + integral + derivative control). We shall postpone further consideration of this point until later.

Higher-order systems

In theory, the transient analysis of higher-order systems can be accomplished in exactly the same manner illustrated for first- and second-order systems. In practice, however, the method is difficult to carry out because it requires solution of a high-degree characteristic equation. For third- and fourth-degree equations, this can still be done analytically but requires considerable labor; for equations of fifth degree or higher, one of the various numerical methods of solution must be used. Computers become increasingly important in the study of such high-order systems.

Although one must be able to determine the roots of the characteristic equation in order to obtain the exact transient response of any particular higher-order system, an understanding of the general nature of all such responses can be obtained in another way. The clue lies in a fundamental theorem of algebra which states that every polynomial $f(x)$ with real coefficients can be expressed as the product of real linear and quadratic factors. This means that the characteristic function of an nth-order linear system can always be represented as the product of n_1 linear factors and n_2 quadratic factors where $n_1 + 2n_2 = n$. Why must we include the quadratic factor? Cannot every such factor be resolved into the product of two linear factors? The answer is that the linear factors must be real, and this will be true only for those quadratic factors which represent critically damped or overdamped second-order systems. Such second-order systems can be resolved into two first-order systems, but underdamped second-order systems whose characteristic equations have complex roots cannot. Stated otherwise, we can say that a higher-order system composed of n first-order lags in cascade cannot oscillate in response to a transient input.

Considered from the above point of view, it is apparent that the transient response of an nth-order system can be regarded as the sum of a number of first- and second-order transients. Assuming that the system is stable (i.e., that the transient response is in fact a transient), the complementary function will always consist of the sum of a number of decaying exponentials and/or exponentially damped sinusoids. In this sense, the higher-order systems introduce nothing new, and the usefulness of a thorough knowledge of first- and second-order systems is thereby extended. However, it must not be forgotten that in order to resolve any particular system into its components and obtain its transient response, one must first determine the roots of the characteristic equation. Moreover, even if one synthesizes a system from known components so that the roots are known, there is still no simple way to obtain a particular transient response or to determine the effects of parameter changes upon transient behavior.

It was largely because of the difficulties involved in applying transient analysis to higher-order systems and in utilizing transient response information in system design and synthesis that frequency analysis became so popular among engineers.

Summary

Transient analysis studies the response of systems to step functions. As the name implies, it is the transient response which is of major interest, the forced response being merely a constant. The transient response of a first-order system is a single decaying exponential characterized by the time constant τ. Second-order transients offer considerably more variety. Depending upon the value of the damping ratio ζ, such a transient will be a nonoscillatory decaying exponential (critical or overdamping, $\zeta \geqq 1$), an exponentially damped sinusoid (underdamping, $0 < \zeta < 1$), or an undamped sinusoid (the harmonic oscillator, $\zeta = 0$). In the latter case, the complementary function is really not a transient at all. The addition of closed-loop proportional control to a second-order system was found to present conflicting requirements for controller gain. On the one hand, high gain is desirable because it minimizes steady-state error, but, on the other hand, it also reduces the damping ratio and so may cause undesirable oscillation. The difficulty may be resolved by adding derivative control to increase damping and integral control to completely eliminate steady-state error. Transient analysis of higher-order systems, although simple enough in theory, is difficult in practice and this is perhaps the major reason for the popularity of frequency analysis, the subject of the next chapter.

CHAPTER 5

Frequency Analysis of Physical Systems

THE steady-state or forced response of stable linear systems to sinusoidal inputs together with the general question of stability constitute the domain of frequency analysis. Such studies are of great utility for a number of reasons. First of all, the analytical mathematical machinery required is particularly simple and may readily be applied to higher-order systems. Thus to establish the steady-state response of a stable system to sinusoidal forcing we need only determine the particular integral, and this of course does not require solution of the characteristic equation. Moreover, it turns out that the behavior of a component of this particular integral (the complex transfer impedance) answers the stability question, again without requiring solution of the characteristic equation. Since the latter is a difficult task for higher-order systems, the advantage of avoiding it is apparent.

A second advantage of frequency analysis lies in the fact that any "well-behaved" periodic function,* as well as many nonperiodic functions, can be represented as the sum of sinusoidal functions by means of a Fourier series or Fourier integral. Now the response of a linear system to n forcings applied simultaneously is the same as the sum of the n responses to each of the forcings applied individually (superposition principle). Hence if we know the response of a system to pure sinusoidal inputs over a range of frequencies, we can easily determine its response to any periodic or nonperiodic function which can be represented by a Fourier series or Fourier integral.

---

* One which satisfies the Dirichlet conditions.

A third advantage of frequency analysis which is of particular interest to control-system engineers lies in the fact that if the frequency responses of individual system components are known, then convenient graphical methods can be used to synthesize open- and closed-loop systems having a particular over-all frequency response.

Special mathematical machinery for frequency analysis

Let us first examine the problem of finding the steady-state response of a stable system to a sinusoidal input. As noted above, we need only determine the particular integral and this does not require solution of the characteristic equation. The task of finding the particular integral becomes particularly simple if we use the complex exponential notation for sinusoidal functions. This will become apparent as we go. Let us begin by considering the following nth-order equation:

$$a_n \frac{d^n y}{dt^n} + a_{n-1} \frac{d^{n-1} y}{dt^{n-1}} + \ldots + a_1 \frac{dy}{dt} + y = F(t), \qquad (5.1)$$

where $F(t)$ is either $y_f \cos \omega_f t$ or $y_f \sin \omega_f t$. We recall the procedure outlined in Chapter 3 for the determination of the particular integral. Table 2 of Chapter 3 tells us that $y_p(t)$ for either a cosine or a sine input is:

$$y_p(t) = B_0 \cos \omega_f t + B_1 \sin \omega_f t, \qquad (5.2)$$

provided that no such undamped harmonic terms appear in the complementary function. Since our system is assumed to be stable, we know that the latter restriction will be satisfied. To determine B_0 and B_1, we would have to substitute (5.2) and its first n derivatives into (5.1), collect sine and cosine terms, and equate their coefficients to y_f and zero, respectively [i.e., if $F(t)$ were $y_f \cos \omega_f t$, the cosine-term coefficient would be equated to y_f and the sine-term coefficient to zero; if $F(t)$ were $y_f \sin \omega_f t$, the reverse would be true]. This would yield two equations whose simultaneous solution would then give us B_0 and B_1 in terms of y_f and the parameters of (5.1). This procedure may become quite laborious if n is large.

The same result can be accomplished much more easily if we use $y_f e^{j\omega_f t}$ for $F(t)$, recalling that $e^{j\omega_f t} = \cos \omega_f t + j \sin \omega_f t$ (Euler formula). Let us see how this simplifies our problem. Since our system is assumed to be stable, $e^{j\omega_f t}$ cannot appear in the complementary function (i.e., $j\omega_f$ cannot be a root of the characteristic equation since pure imaginary roots are excluded). Hence, according to Table 2, our particular integral is:

$$y_p(t) = Ae^{j\omega_f t}. \tag{5.3}$$

It is now a simple matter to evaluate A. Once more we begin by substituting (5.3) and its first n derivatives into (5.1). But the first derivative of $Ae^{j\omega_f t}$ is just $Aj\omega_f e^{j\omega_f t}$, the second is $A(j\omega_f)^2 e^{j\omega_f t}$, and the nth is $A(j\omega_f)^n e^{j\omega_f t}$. Hence the expression obtained by this substitution can be written down immediately:

$$A[a_n(j\omega_f)^n + a_{n-1}(j\omega_f)^{n-1} + \ldots + a_1(j\omega_f) + 1]e^{j\omega_f t} = y_f e^{j\omega_f t}. \tag{5.4}$$

We at once recognize that the bracketed term is nothing more than the characteristic function of the system with $j\omega_f$ as the variable. Let us define this algebraic polynomial in $j\omega_f$ as the transfer impedance, Z. It is now obvious that $A = y_f/Z$, and our desired solution is simply:

$$y_p(t) = \frac{y_f}{Z} e^{j\omega_f t} = \frac{y_f}{Z} (\cos \omega_f t + j \sin \omega_f t). \tag{5.5}$$

It only remains to translate (5.5) into more familiar form. To do so, we must probe a bit more deeply into the realm of complex numbers for, in general, the transfer impedance Z will be such a number. Let us therefore write it in the following form:

$$Z = \alpha + j\omega = \Re(Z) + j\Im(Z) \tag{5.6}$$

and recall that the real number α is defined as the real part of Z, $\Re(Z)$, and the real number ω as the imaginary part of Z, $\Im(Z)$. The impedance Z may be represented geometrically in the complex plane either in terms of the rectangular coordinates (α, ω) or the polar coordinates (Z_0, ϕ). Both representations are shown in Fig. 34. From the figure, it is apparent that:

$$\alpha = Z_0 \cos \phi \tag{5.7}$$
$$\omega = Z_0 \sin \phi, \tag{5.8}$$

where $Z_0 = (\alpha^2 + \omega^2)^{1/2}$ and is called the absolute value of the transfer impedance $(Z_0 \equiv |Z|)$. Substituting (5.7) and (5.8) in (5.6) and applying the Euler formula, we have:

$$Z = Z_0 (\cos \phi + j \sin \phi) = Z_0 e^{j\phi}. \qquad (5.9)$$

Finally, substituting (5.9) into (5.5), we obtain the final form of our solution:

$$y_p(t) = \frac{y_f e^{j\omega_f t}}{Z_0 e^{j\phi}} = \frac{y_f}{Z_0} e^{j(\omega_f t - \phi)} = \frac{y_f}{Z_0} [\cos(\omega_f t - \phi) + j \sin(\omega_f t - \phi)].$$

$$(5.10)$$

If y_f and the parameters of (5.1) are real (as they always will be for an actual physical system), the real part of (5.10) is the

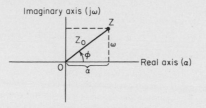

FIG. 34. GEOMETRIC REPRESENTATION OF COMPLEX NUMBER Z

steady-state solution of (5.1) for $F(t) = y_f \cos \omega_f t$, and the imaginary part of (5.10) is the solution for $F(t) = y_f \sin \omega_f t$. Equation (5.10) thus defines a sinusoid of frequency ω_f and amplitude y_f/Z_0 which is shifted ϕ radians along the time axis in relation to the input sinusoid. Since the amplitude of the output sinusoid is y_f/Z_0, and that of the input sinusoid is y_f, their ratio, which we can call the system gain, is just $1/Z_0$. The angle ϕ is called the phase angle or simply the phase. It is best defined in terms of the two equations (5.7) and (5.8) but is more commonly defined as $\tan^{-1}[\mathcal{I}(Z)/\mathcal{R}(Z)]$. This is all right provided one is careful to locate the proper quadrant in which ϕ lies. We note that a positive value of ϕ denotes a phase lag (i.e., the output lags behind the input) and that a negative value denotes a phase lead.

We have thus discovered that the steady-state response of a stable linear system to a sinusoidal input is a sinusoid of the same frequency but of different amplitude and phase. We have further discovered that all of the relevant information about the response is contained in the complex transfer impedance Z which is nothing more than the characteristic function of the system with $(j\omega_f)$ as the variable. The complex quantity Z may be regarded as a vector whose absolute value Z_0 is the reciprocal of the system gain and whose direction ϕ defines the phase difference between input and output sinusoids. Both gain and phase are functions of the forcing frequency ω_f and these two frequency functions (gain characteristic and phase characteristic) completely define the response of the system to sinusoidal forcing.

What must be the nature of the gain and phase characteristics if an arbitrary periodic input wave form is to be reproduced by the output without distortion? First, the gain must be the same at all frequencies (the gain characteristic must be "flat") so that each Fourier frequency component of the input will appear at the output with the proper relative amplitude. Second, the phase characteristic must either be zero at all frequencies or else must be directly proportional to frequency so that all Fourier input components will "arrive" simultaneously at the output. In the first case, there will be neither distortion nor delay; in the second, there will be delay but no distortion. Distortion caused by an improper gain characteristic is called gain distortion and that due to an incorrect phase characteristic is called phase distortion.

In Laplace transform terminology, it is apparent that the system transfer function with $(j\omega_f)$ substituted for s [the frequency transfer function $G(j\omega_f)$] is just the reciprocal of the complex transfer impedance Z. Similarly, the absolute value of $G(j\omega_f)$, usually designated by $|G(j\omega_f)|$, is the reciprocal of Z_0 and thus directly gives the system gain. Hence we can write (5.5) as:

$$y_p(t) = y_f G(j\omega_f)e^{j\omega_f t} = y_f G(j\omega_f)[\cos \omega_f t + j \sin \omega_f t] \quad (5.11)$$

and (5.10) as:

$$y_p(t) = y_f |G(j\omega_f)| e^{j(\omega_f t + \phi')}$$
$$= y_f |G(j\omega_f)|[\cos (\omega_f t + \phi') + j \sin (\omega_f t + \phi')], \quad (5.12)$$

where:

$$\phi' = \tan^{-1}\frac{\mathcal{J}G(j\omega_f)}{\mathcal{R}G(j\omega_f)} = -\phi. \tag{5.13}$$

Note that a positive value of the phase angle ϕ' defined in (5.13) now represents a phase lead and a negative value a phase lag. Thus $\phi' = -\phi$. It is then apparent that the classical and Laplace transform representations are very similar and equally convenient. The latter has the advantage that $|G(j\omega_f)|$ gives system gain directly whereas Z_0 gives the reciprocal of the gain. On the other hand, it is usually easier to resolve Z into its real and imaginary components. As we shall see, both representations are commonly used.

Stability

So far we have simply assumed that our system is stable, but now we must "pull our head from the sand" and ask how we determine whether this assumption is a valid one. Recalling our discussion in Chapter 3, we know that a system will be stable if its characteristic equation has no roots on the imaginary axis or in the right half of the s-plane. The presence of such roots is easily detected for first- and second-order systems by direct solution of the characteristic equation, but this approach rapidly becomes impractical for higher-order systems. Fortunately, it turns out that the same complex transfer impedance which appears in the particular integral of stable systems also provides an answer to the general question of stability. The vector properties of complex numbers provide the basis for a graphical solution of the problem. Let us see how this works out.

Let us write the characteristic function (or complex transfer impedance) of an nth-order linear system as follows:

$$Z(s) = a_n s^n + a_{n-1} s^{n-1} + \ldots + a_1 s + 1, \tag{5.14}$$

where s is the complex variable $\alpha + j\omega$. According to the fundamental theorem of algebra, (5.14) can be written as the product of n linear factors:

$$Z(s) = a_n(s - s_1)(s - s_2) \ldots (s - s_n), \tag{5.15}$$

where s_1, s_2, . . . s_n are the n roots* of the characteristic equation $Z(s) = 0$. Since substitution of any one of these roots for s in (5.14) or (5.15) yields $Z(s) = 0$, the roots can also be called the zeros of the impedance.

Now let us direct our attention to two related complex planes. The first of these is the s-plane in which we plot the locus of the complex variable s. The second is the Z-plane in which we plot the locus of $Z(s)$ defined by Eq. (5.14) as s varies in the s-plane. The plot of $Z(s)$ is called a conformal map of the plot of s. This is illustrated in Fig. 35 where we have allowed s to

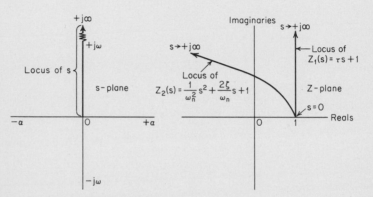

FIG. 35. CONFORMAL MAPPING OF s-PLANE INTO THE Z-PLANE

vary from zero to $+j\infty$ along the positive imaginary axis of the s-plane and have plotted the corresponding loci for two particular forms of $Z(s)$ in the Z-plane. Each point on the locus of s determines a corresponding point on the locus of $Z_1(s)$ defined by $Z_1(s) = \tau s + 1$ and on the locus of $Z_2(s)$ defined by $Z_2(s) = \dfrac{1}{\omega_n{}^2} s^2 + \dfrac{2\zeta}{\omega_n} s + 1$. Let us now consider the behavior of a single factor of $Z(s)$, say $(s - s_1)$, in the s-plane as s varies around an arbitrary closed contour in this plane. In Fig. 36, we have represented the vector $(s - s_1)$ for three different values of s_1. In one case s_1 lies outside of the closed s-contour, in a second it lies within the contour, and in a third it lies upon the contour. Suppose we now allow s to make one complete counter-

---

* Some or all of these roots may be equal.

clockwise trip around the closed contour and examine the behavior of $(s - s_1)$. From the figure it becomes obvious that if
s_1 lies outside of the contour, the net rotation (or change in
phase) of the vector $(s - s_1)$ will be zero. On the contrary, if s_1
lies within the contour, $(s - s_1)$ will undergo one complete

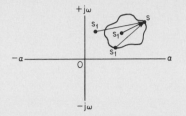

FIG. 36. BEHAVIOR OF $(s - s_1)$ FOR CLOSED LOCUS OF s

counterclockwise rotation, i.e., it will show a net phase change
of $+2\pi$ radians or $+360°$. Finally, if s_1 lies on the contour, the
vector $(s - s_1)$ will vanish (i.e., equal zero) when $s = s_1$, and
will undergo an abrupt phase change of $+180°$ at this point.
This behavior provides a basis for answering the stability question.

To do so, let us first define a stability test contour in the s-plane

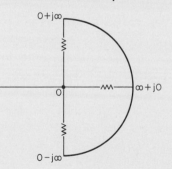

FIG. 37. STABILITY TEST CONTOUR IN THE s-PLANE

shown in Fig. 37. It is bounded by the imaginary axis extending
from $-j\infty$ to $+j\infty$ and by a semicircular arc of infinite radius
which encloses the entire right half-plane. Now let us examine
the behavior of $Z(s)$ in Eq. (5.15) as s makes one complete
counterclockwise trip around our s-test contour. For every root

of $Z(s) = 0$ which lies in the right half of the s-plane, the vector $Z(s)$ will rotate once counterclockwise around the origin of the Z-plane. Thus, if there are k such roots, $Z(s)$ will increase in phase by $k(2\pi)$ radians.* Finally, for every conjugate pair of pure imaginary roots, $Z(s)$ will vanish, i.e., the locus of $Z(s)$ will pass through the origin of the Z-plane. Thus, if the system is stable, the locus of $Z(s)$ in the Z-plane will not pass through the origin and will have a net phase change of zero.

In practice, the test is most often applied by plotting only a portion of the conformal locus on the Z-plane. For example it is apparent from our previous discussion that as s goes from zero to $+j\infty$ along the imaginary axis of Fig. 37, $Z(s)$ will increase in phase by $\pi/2$ for every root of $Z(s) = 0$ in the left half-plane,† and decrease by $\pi/2$ for every root of $Z(s) = 0$ in the right half-plane. The net change in phase ϕ is then: $\phi = (n_l - n_r)\pi/2$ where n_l is the number of roots in the left half-plane and n_r the number in the right half-plane. Assuming no roots to be on the imaginary axis, $n_l = n - n_r$ where n is the order of the system so that $\phi = (n - 2n_r)\pi/2$. We can summarize this neatly by saying that if an nth-order system is to be stable, the locus of $Z(s)$ must rotate n quadrants in the positive direction as s goes from 0 to $(0 + j\infty)$. Stated from the opposite viewpoint we can say that the number of roots in the right half-plane n_r is $[(n/2) - (\phi/\pi)]$. If there are any pure imaginary roots, these relationships no longer hold for $n \neq n_l + n_r$. The presence of such roots is revealed by the passage of $Z(s)$ through the origin of the Z-plane and this in itself indicates instability.

Examples of such loci for stable and unstable systems are shown in Fig. 38. Note that $Z(s) = 1$ when $s = 0$ in all cases which is also apparent from Eq. (5.14).

Finally, let us note that the locus of $Z(s)$ as s goes from 0 to $+j\infty$ along the imaginary axis is exactly the locus of the complex transfer impedance which appears in Eq. (5.5). Thus this single plot (which we shall later come to call a Nyquist diagram) not

---

* Note that multiplication of the vectors $(s - s_1)(s - s_2) \ldots (s - s_n)$ in Eq. (5.15) involves multiplication of their absolute values but *addition* of their phase angles.

† Strictly speaking, $Z(s)$ increases in phase by π for every pair of conjugate complex roots and by $\pi/2$ for every real root in the left half-plane.

only defines gain and phase functions for stable systems but also tells whether or not the system is in fact stable. Some interesting questions arise in this regard.

Suppose, for example, that $Z(s)$ vanished (i.e., passed through the origin) for some particular value of $j\omega_f$. Should we then proceed to write Eq. (5.5) as:

$$y_p(t) = \frac{y_f}{0} e^{j\omega_f t} = \infty e^{j\omega_f t} \qquad (5.16)$$

and claim that (5.16) is a solution of (5.1)? In the words of a professor of mathematics (Agnew*) to do so would be "one of

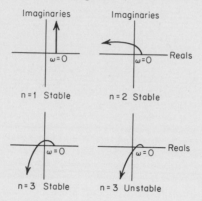

FIG. 38. $Z(s)$ LOCI FOR STABLE AND UNSTABLE SYSTEMS

Based on J.D. Trimmer, *Response of Physical Systems* (New York, John Wiley and Sons, Inc., 1950).

those inglorious performances which cannot be tolerated in a rational society," and we would certainly not wish to be guilty of that! We hasten to point out, therefore, that if $j\omega_f$ is a zero of the impedance (or a root of the characteristic equation), then (5.3) is not the proper form for the particular integral. Instead it is:

$$y_p(t) = Ate^{j\omega_f t} \qquad (5.17)$$

and this describes a harmonic oscillation whose amplitude increases linearly with time to approach infinity as t approaches

* R. P. Agnew, *Differential Equations* (New York, McGraw-Hill Book Company, Inc., 1942).

infinity. Hence, although it might be "inglorious," it is perhaps not completely impractical to accept the implications of (5.16)!

We might also point out here that if a system is unstable because of the presence of pure imaginary roots of multiplicity one, *growing* harmonic oscillations arise only from the *forced* response to "resonant" sinusoidal inputs. On the other hand, if there are roots in the right half-plane or roots of multiplicity greater than one on the imaginary axis, then growing oscillations arise from the transient response or complementary function. We shall have more to say about the problem of stability in connection with the particular examples of frequency analysis which we are now ready to examine.

A first-order system

Using the complex exponential notation, the differential equation for sinusoidal forcing is:

$$\tau \dot{y} + y = y_f e^{j\omega_f t}. \tag{5.18}$$

The complex transfer impedance Z or the reciprocal of the frequency transfer function $1/G(j\omega_f)$ is obtained by substituting $j\omega_f$ into the characteristic function of the system and rearranging to conform to complex number notation:

$$Z = \frac{1}{G(j\omega_f)} = 1 + j(\tau\omega_f). \tag{5.19}$$

The gain characteristic $1/Z_0$ or $|G(j\omega_f)|$ and the phase characteristic ϕ or ϕ' can now be written at once:

$$\frac{1}{Z_0} = |G(j\omega_f)| = \frac{1}{(1 + \tau^2\omega_f^2)^{1/2}} \tag{5.20}$$

$$\phi = \tan^{-1} \tau\omega_f = -\phi'. \tag{5.21}$$

Equations (5.20) and (5.21) completely define the steady-state response of a first-order system to sinusoidal forcing.

It is customary to display these functions graphically in one of two ways. The first of these is a polar plot in the complex plane of the frequency transfer function $G(j\omega_f)$ or its reciprocal, the complex transfer impedance Z. Such a polar plot is called a Nyquist diagram. For each value of ω_f, a point is plotted whose polar coordinates will be $[|G(j\omega_f)|, \phi']$ or (Z_0, ϕ). The curve obtained by joining these points constitutes the Nyquist diagram. For the first-order system under consideration, the polar plot of $G(j\omega_f)$ is shown in Fig. 39 and that for Z in Fig. 40. The figures tell us that when $\omega_f = 0$, the gain is 1 and the phase is zero.

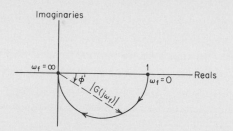

FIG. 39. NYQUIST DIAGRAM OF FIRST-ORDER G

As ω_f increases, the gain progressively decreases and the phase lag increases to approach zero and $\pi/2$ (90°), respectively, as ω_f approaches infinity.

We can also examine Figs. 39 and 40 from the standpoint of stability. In Fig. 40, we see that the $Z(s)$ locus does not pass

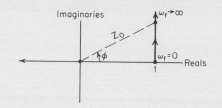

FIG. 40. NYQUIST DIAGRAM OF FIRST-ORDER Z

through the origin and that it rotates one quadrant in the positive direction (i.e., $+\pi/2$ or $+90°$) as $j\omega_f$ goes from 0 to $+j\infty$. Thus our system is stable. What about the plot of the reciprocal locus $G(j\omega_f)$ in Fig. 39? The stability requirement here is that $G(j\omega_f)$ rotate $-\pi/2$ or $-90°$ and not have any "infinities" as $j\omega_f$ goes from 0 to $+j\infty$, and once more the criteria are satisfied. Of course, for this first-order system it is a simple matter to obtain the single root of the characteristic equation directly. It is obviously $-1/\tau$, and since τ is a positive real number, we know our system must be stable. Finally, let us summarize the relationship between the Nyquist diagrams of any function, say $G(j\omega_f)$, and that of its reciprocal $1/G(j\omega_f)$. It is quite simple. The absolute value (or vector length) of $1/G(j\omega_f)$ is the reciprocal of that of $G(j\omega_f)$ and the phase ϕ of $1/G(j\omega_f)$ is the negative of the phase ϕ' of $G(j\omega_f)$.

The second form of graphical display of the frequency response is a plot of phase ϕ' and [20 log (gain)] against log ω_f. The quantity [20 log (gain)] expresses the gain in decibels (db), and the vertical gain scale becomes a linear scale in decibel units. The horizontal frequency axis may be nondimensionalized which, in the present case, is accomplished by plotting log $(\tau\omega_f)$ instead

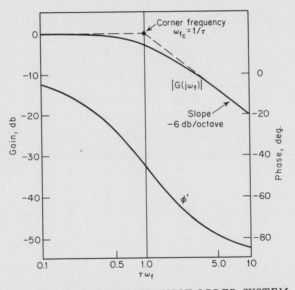

FIG. 41. BODE DIAGRAM, FIRST-ORDER SYSTEM

of log (ω_f). Such a plot is called a Bode diagram, and its appearance for our first-order system is shown in Fig. 41.

Looking first at the gain curve, we note that at low forcing frequencies ($\tau\omega_f <$ about 0.3), it is flat at zero decibels. As $\tau\omega_f$ increases, the gain curve begins to fall, reaching -3 db at $\tau\omega_f = 1$, and at higher frequencies becoming asymptotic to a line having a slope of -6 db/octave or -20 db/decade.* The curve may be approximated by its high- and low-frequency asymptotes which intersect at the "corner" frequency $\omega_{f_c} = 1/\tau$. In engineering jargon, we might say that "the frequency response is essentially flat (i.e., ± 3 db) to ω_{f_c} and attenuates 6 db/octave at higher frequencies." Looking next at the phase curve, we see that it asymptotically approaches zero at low frequencies and $-90°$ at high frequencies, and that it is $-45°$ at ω_{f_c}.

Just as in the case of the Nyquist diagram, we can also plot the reciprocal function $1/G(j\omega_f)$ as a Bode diagram. Because the absolute values (or vector lengths) in this plot are expressed in logarithmic units, the gain of $1/G(j\omega_f)$ is just the negative of that of $G(j\omega_f)$. Hence the Bode plot of $1/G(j\omega_f)$ is just that of $G(j\omega_f)$ with the signs changed for both gain and phase.

The inverse problem for the first-order system is easily solved from a knowledge of either the gain or the phase characteristic. We simply evaluate $\tau\omega_f$ from Fig. 41, Eq. (5.20) or (5.21), and, since we know ω_f, we readily find τ.

A second-order system

The differential equation for sinusoidal forcing of a second-order system is:

$$\frac{1}{\omega_n{}^2}\ddot{y} + \frac{2\zeta}{\omega_n}\dot{y} + y = y_f e^{j\omega_f t}. \tag{5.22}$$

The complex transfer impedance, or the reciprocal of the frequency transfer function is readily seen to be:

$$Z = \frac{1}{G(j\omega_f)} = \frac{1}{\omega_n{}^2}(j\omega_f)^2 + \frac{2\zeta}{\omega_n}(j\omega_f) + 1. \tag{5.23}$$

Remembering that $j^2 = -1$, this can be written:

$$Z = \frac{1}{G(j\omega_f)} = -\left(\frac{\omega_f}{\omega_n}\right)^2 + j2\zeta\left(\frac{\omega_f}{\omega_n}\right) + 1 \tag{5.24}$$

---

* An octave is an interval in which frequency is doubled; a decade is an interval in which frequency is increased tenfold.

and using β to denote the dimensionless ratio of forcing frequency to natural frequency, i.e., $\beta \equiv \omega_f/\omega_n$, we have finally:

$$Z = \frac{1}{G(j\omega_f)} = (1 - \beta^2) + j2\zeta\beta. \tag{5.25}$$

The gain characteristic, $1/Z_0$ or $|G(j\omega_f)|$, and the phase characteristic, ϕ or ϕ', can now be written easily:

$$\frac{1}{Z_0} = |G(j\omega_f)| = \frac{1}{[(1 - \beta^2)^2 + (2\zeta\beta)^2]^{1/2}} \tag{5.26}$$

$$\phi = \tan^{-1}\left(\frac{2\zeta\beta}{1 - \beta^2}\right) = -\phi'. \tag{5.27}$$

Equations (5.26) and (5.27) completely define the steady-state response of a second-order system to sinusoidal forcing. Again, we may display them in the form of Nyquist or Bode diagrams.

The Nyquist diagram for the frequency transfer function $G(j\omega_f)$ is shown in Fig. 42, and that for the transfer impedance

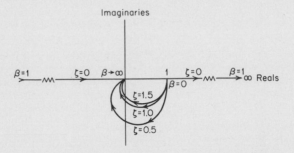

FIG. 42. NYQUIST DIAGRAM OF SECOND-ORDER G

Z in Fig. 43. The several curves in each figure correspond to different values of the damping ratio ζ. Note the special case of the harmonic oscillator ($\zeta = 0$) where, at $\beta = 1$, the gain becomes infinite,* and the phase angle ϕ' "jumps" from $0°$ to $-180°$.

Once more we can examine these figures from the standpoint of stability. For the $Z(s)$ plot in Fig. 43, we see that for $\zeta > 0$ the locus does not pass through the origin and rotates two quadrants counterclockwise ($+\pi$ or $+180°$) as $j\omega_f$ goes from 0 to

* With the reservation previously noted in the section on stability.

$+j\infty$, thus satisfying our stability criteria. However, for $\zeta = 0$, $Z(s)$ passes through the origin and "flips" $+180°$ at $\beta = 1$. This implies a conjugate pair of pure imaginary roots and thus instability. In the plot of $G(j\omega)$ in Fig. 42, the stability criteria of $-180°$ rotation and no infinities are met for $\zeta > 0$, but when

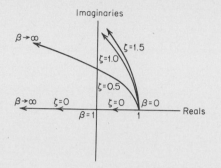

FIG. 43. NYQUIST DIAGRAM OF SECOND-ORDER Z

$\zeta = 0$, $G(s)$ has an infinite discontinuity and flips $-180°$ at $\beta = 1$. A more direct approach to the stability problem is, of course, still practical for a second-order system, for the quadratic characteristic equation can readily be solved.

The gain and phase characteristics are perhaps more easily visualized in the Bode diagram of Figs. 44a and 44b. Looking first at the gain curves, we note that for values of $\zeta < 1/(2)^{1/2}$, these curves have maxima or resonance peaks which occur at $\beta = (1 - 2\zeta^2)^{1/2}$ and have values equal to $1/[2\zeta(1 - \zeta)^{1/2}]$. Thus, in the special case of the harmonic oscillator ($\zeta = 0$), $|G(j\omega_f)|$ approaches infinity at $\beta = 1$, i.e., when the forcing frequency equals the natural frequency. As ζ increases, the maximum gain occurs at progressively lower values of β, reaching $\beta = 0$ at $\zeta = 1/(2)^{1/2}$. For large values of β, the gain curves all approach a straight line having a slope of -12 db/octave or -40 db/decade. This high-frequency asymptote intersects the zero-decibel ordinate at an abscissa value corresponding to $\beta = 1$, at which point $\omega_f = \omega_n$. Looking next at the phase curves, we note (1) that they all approach zero at low frequencies and $-180°$ at high frequencies, and (2) that $\phi' = -90°$ at

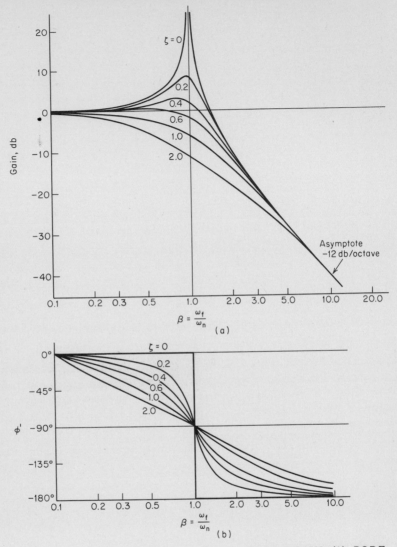

FIG. 44. (a) BODE DIAGRAM, SECOND-ORDER GAIN; (b) BODE DIAGRAM, SECOND-ORDER PHASE

$\beta = 1$ for all values of ζ. The slope $d\phi'/d\beta$ is equal to $1/\zeta$ at $\beta = 1$. We also note the jump in ϕ' from $0°$ to $-180°$ at $\beta = 1$ for the harmonic oscillator.

There are several ways of solving the inverse problem. For example, we can evaluate ω_n directly by finding the value of

ω_f at which the high-frequency-gain asymptote intersects the zero-db ordinate, or at which $\phi' = -90°$. The phase slope $d\phi'/d\beta$ at this point then yields $1/\zeta$. It is also possible to calculate both ω_n and ζ if we know both the gain and the phase at a single forcing frequency, or if we know either the gain or the phase at each of two frequencies. For example, suppose that we know $|G(j\omega_f)|$ and ϕ' at a particular value of ω_f. It can then be shown that:

$$\beta = \left[1 - \frac{\cos \phi'}{|G(j\omega_f)|} \right]^{1/2} \qquad (5.28)$$

and
$$\zeta = \frac{-\sin \phi'}{2\beta |G(j\omega_f)|}. \qquad (5.29)$$

Equations (5.28) and (5.29) follow from the fact that the cartesian coordinates of the point $G(j\omega_f)$ in the complex plane are $(|G(j\omega_f)|^2(1 - \beta^2), -j|G(j\omega_f)|^2 2\zeta\beta)$. We shall not go into the details of obtaining ω_n and ζ from other combinations of values for gain and phase.

Higher-order systems

One of the major advantages of frequency analysis is the ease with which it may be applied to higher-order systems. Thus the transfer impedance or frequency transfer function is easily written directly from the differential equation, the gain and phase characteristics derived therefrom, and graphic stability tests applied without ever having to solve the characteristic equation. Solution of the inverse problem involves simply a generalization of the method applied to the second-order system just considered. Gain and/or phase values at the required number of frequencies are put into a set of simultaneous equations whose solution yields the n-parameters.

The advantages of the frequency method in dealing with higher-order systems extends to their synthesis from simpler components. Thus we noted in Chapter 4 that there is no simple way of combining the known transient responses of the components to obtain the particular transient response of the combination. This is not true of the frequency response. In the steady state, the input to every component block is a sinusoid of the same frequency as the forcing frequency. Each block merely changes the amplitude and phase of this sinusoid. Hence all we have to do to obtain the over-all frequency response of the cascade combination is multiply the gain characteristics and

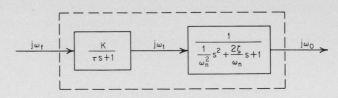

FIG. 45. FIRST- AND SECOND-ORDER SYSTEMS IN CASCADE

sum the phase characteristics of the component blocks. Moreover, both of these operations can be conveniently performed graphically, and the sequence of particular blocks in the cascade combination is immaterial.

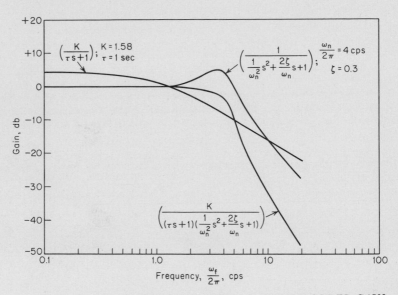

FIG. 46. BODE DIAGRAM SYNTHESIS OF THIRD-ORDER GAIN

To illustrate the ease of synthesis using the frequency approach, let us build a third-order system by connecting a first-order and a second-order system in cascade (Fig. 45). Graphic synthesis is particularly simple if we use the logarithmic Bode diagram, for now the multiplication of gains can be accomplished simply by addition of decibels. Thus in Fig. 46 we have plotted the gain and in Fig. 47 the phase characteristics of the two component blocks and, by merely adding their ordinates, we obtain the corresponding characteristics for the third-order

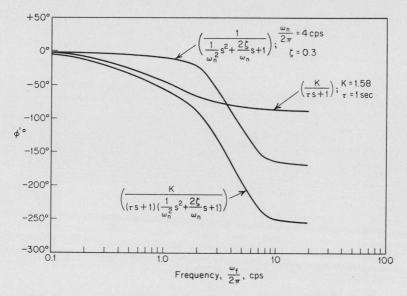

FIG. 47. BODE DIAGRAM SYNTHESIS OF THIRD-ORDER PHASE

system of Fig. 45. We can apply the same procedure to any number of such blocks in cascade and, as noted above, their sequence is immaterial. Having synthesized the frequency response of the higher-order system, we can use the Nyquist diagram to test its stability. It is therefore apparent that the frequency approach to synthesis is a particularly simple one. In the next section, we shall explore some of its special advantages in dealing with feedback control systems.

Special applications of frequency analysis to feedback systems

Any single-loop feedback system employing any of the control modes discussed in Chapter 4 can always be represented as an equivalent unity feedback system whose block diagram, using

Laplace-transform notation, is shown in Fig. 48. The transfer function $G(s)$ of the single block now represents a combination of the transfer functions of the controlled system and of all controlling-system operations except the error-detector function $y_e(s) = y_i(s) - y_o(s)$. The latter corresponds to the operation of

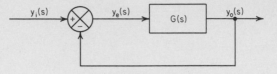

FIG. 48. SINGLE-LOOP UNITY FEEDBACK SYSTEM

closing the loop of a unity-gain proportional control system around the composite block $G(s)$. Recalling the discussion in Chapter 2, we can at once write three related transfer functions for such a system.* They are:

$$\frac{y_o(s)}{y_e(s)} = G(s) \qquad \text{(open-loop transfer function)} \qquad (5.30)$$

$$\frac{y_o(s)}{y_i(s)} = \frac{G(s)}{1 + G(s)} \qquad \text{(closed-loop transfer function)} \qquad (5.31)$$

and

$$\frac{y_e(s)}{y_i(s)} = \frac{1}{1 + G(s)} \qquad \text{(error transfer function).} \qquad (5.32)$$

We would now like to consider the following two problems: (1) the synthesis of the open-loop frequency response $G(j\omega_f)$ from its component blocks, and (2) the derivation of the corresponding closed-loop frequency response together with the assessment of closed-loop stability.

The controlled-system component of $G(j\omega_f)$ can be readily synthesized from first- and second-order lags using the Bode graphical method already described, but how do we include the controlling-system components in $G(j\omega_f)$? A general controlling-system block diagram including the three conventional

---

* For present purposes, we have included the constant k_1 of Chapter 2 along with all other system constants in $G(s)$. Any dynamic elements in the feedback loop can also be included in $G(s)$.

control modes appears in Fig. 49. The top block represents proportional control action, the middle block derivative control, and the bottom block integral control. The only notation that we have not previously encountered is the use of $(1/s)y_e(s)$ to represent the Laplace transform of the integral $\int y_e(t)\, dt$.

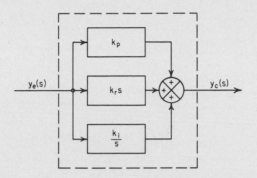

FIG. 49. PROPORTIONAL + DERIVATIVE + INTEGRAL CONTROLLER

Now it is actually very simple to obtain the frequency response of each of these blocks. Let us assume that the input to each block is the unit sinusoid, $\sin \omega_f t$. Then the proportional block merely multiplies this input by the constant k_p, i.e., it produces a gain of k_p and a phase difference of zero for all values of ω_f. The output of the derivative block is just $k_r[d(\sin \omega_f t)/dt]$, and this is $k_r\omega_f \cos \omega_f t$. This means that the derivative block produces a gain proportional to ω_f and a phase lead of $\pi/2$ or 90° at all frequencies. Finally, the output of the integral block is just $k_i \int \sin \omega_f t\, dt$ and this is $-(k_i/\omega_f) \cos \omega_f t$. This means that the integral block produces a gain which is inversely proportional to frequency and a phase lag of $\pi/2$ or 90° at all frequencies.

Bode plots for these three blocks are shown in Fig. 50. Note that the derivative and integral constants k_r and k_i only shift the position of the gain curves but do not change their slopes. It is now a simple matter to incorporate these controlling blocks into the over-all transfer function $G(s)$ by adding them to the controlled-system components in the Bode diagram remembering, however, that the three control mode blocks are connected in parallel. In terms of the curves of Fig. 50, we can say that derivative control is used to increase high-frequency gain while simultaneously improving stability by introducing a phase lead. On the contrary, integral control is used to improve the low-frequency response, in particular to eliminate steady-state error

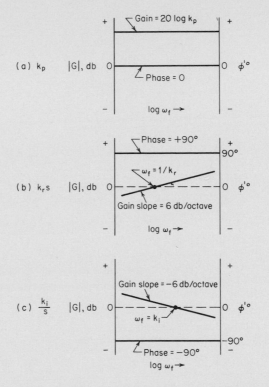

*FIG. 50. BODE DIAGRAMS FOR VARIOUS CONTROL MODES;
(a) PROPORTIONAL; (b) DERIVATIVE; (c) INTEGRAL*

by introducing "infinite" gain at $\omega_f = 0$. However, because it introduces additional phase lag, it may cause instability. The question of the relationship of the gain and phase characteristics of the composite open-loop transfer function $G(s)$ to closed-loop performance and stability now brings us to our second problem.

If we know the open-loop frequency transfer function $G(j\omega_f)$, then there are graphical methods available for synthesizing the closed-loop frequency transfer function $G(j\omega_f)/[1 + G(j\omega_f)]$. The first of these makes use of the Nyquist diagram of $G(j\omega)$ (the transfer locus in the G-plane). Such a plot is shown in Fig. 51. A moment's reflection will show that with reference to

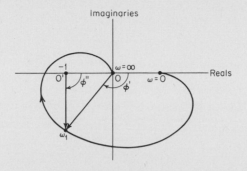

FIG. 51. NYQUIST SYNTHESIS OF CLOSED-LOOP FREQUENCY RESPONSE

the point $-1 + j0$ as origin, this same plot represents the locus of $[1 + G(j\omega_f)]$. Hence, at any frequency on the locus, say ω_1, the length of the vector $0\omega_1$ is the gain and the angle ϕ' is the phase of $G(j\omega_f)$, while the length of the vector $0'\omega_1$ is the gain and the angle ϕ'' is the phase of $1 + G(j\omega_f)$. Then, since the closed-loop frequency transfer function is $G(j\omega_f)/[1 + G(j\omega_f)]$, the quotient $0\omega_1/0'\omega_1$ will give the closed-loop gain and the difference $\phi' - \phi''$ will give the closed-loop phase at ω_1. Hence, using a compass and a pair of dividers, the closed-loop locus can be easily constructed. Once obtained, it may be transferred to a Bode diagram if desired.

The synthesis just described can be simplified even further by superimposing contours of constant closed-loop gain and constant closed-loop phase upon the transfer locus plot. Thus it turns out that the locus of any constant closed-loop gain M is a circle of radius $|M/(M^2 - 1)|$, with its center located on the real axis at $\alpha_c = -M^2/(M^2 - 1)$. In like fashion, the locus for any constant closed-loop phase N is a circle of radius $(1/2N)(N^2 + 1)^{1/2}$ with its center at the point $\alpha_c = -\frac{1}{2}$; $\omega_c = +1/2N$. A number of these "M-circles" are shown in Fig. 52 and the corresponding "N-circles" in Fig. 53. If these are superimposed on the plot of the transfer locus $G(j\omega_f)$, the closed-loop gain and phase at any given ω_f can be read off directly. In relation to the M-circles, it is worth noting that a gain of one is represented by a straight line parallel to the imaginary axis at $\alpha = -0.5$ and that the

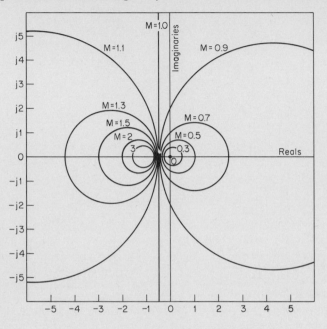

FIG. 52. M CIRCLES FOR NYQUIST G PLOT

Calculated according to G. S. Brown and D. P. Campbell, *Principles of Servomechanisms* (New York, John Wiley and Sons, Inc., 1948).

region to the right of this line represents all gains less than one, while the region to the left represents all gains greater than one.

The same synthesis can be performed using the reciprocal locus $1/G(j\omega_f)$. In this case, the M-contours become concentric circles centered at the point $\alpha_c = -1$; $\omega_c = 0$, with radius equal to $1/M$, while the N-contours become straight lines of slope $-N$ passing through the point $\alpha_c = -1$; $\omega_c = 0$. These are shown in Fig. 54.

Finally, the M and N contours can be transformed so that they can be superimposed on a plot of the open-loop gain in decibels against the open-loop phase angle. This form of plot, shown in Fig. 55, is called a Nichols diagram. If the open-loop Bode plot is transferred to the Nichols diagram, as shown in the figure, the closed-loop frequency response can be readily determined.

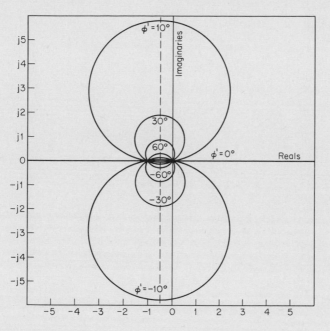

FIG. 53. N CIRCLES FOR NYQUIST G PLOT

Calculated according to G. S. Brown and D. P. Campbell, *Principles of Servomechanisms* (New York, John Wiley and Sons, Inc., 1948).

We have thus discovered that there are several different graphical methods for determining the closed-loop frequency response of a unity feedback system from a knowledge of the open-loop response $G(j\omega_f)$. Which method is chosen is largely a matter of individual preference and experience. Let us now consider our next problem, i.e., the assessment of closed-loop stability from a knowledge of the open-loop response.

It turns out that this problem can be solved by working directly with the open-loop response, i.e., we do not have to synthesize the closed-loop response at all. Let us see how this comes about. We know that the closed-loop system will be stable if none of the roots of its characteristic equation lie on the imaginary axis or in the right half-plane. But what is the characteristic equation of the closed-loop system? We can best answer this by beginning with the differential equation of the

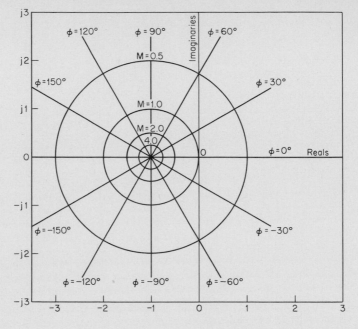

FIG. 54. M AND N CONTOURS FOR NYQUIST Z PLOT

Calculated according to G. S. Brown and D. P. Campbell, *Principles of Servomechanisms* (New York, John Wiley and Sons, Inc., 1948).

open-loop system written in operator form, i.e., we let s stand for the operation of differentiation as we did in Chapter 2:

$$Z(s)y = N(s)y_e. \tag{5.33}$$

Both $Z(s)$ and $N(s)$ represent rational algebraic polynomials in the differential operator s. We have included $N(s)$ in the forcing term in order to include all possible control modes, for, as we have seen in Chapter 4, some of these introduce derivative terms into the error-forcing function. We assume that $N(s)y_e$ is always finite. We note that the open-loop characteristic equation is $Z(s) = 0$ where s is now an algebraic quantity, and that the open-loop transfer function $G(s) = N(s)/Z(s)$ where s is now the complex Laplace transform variable.* If we close a unity

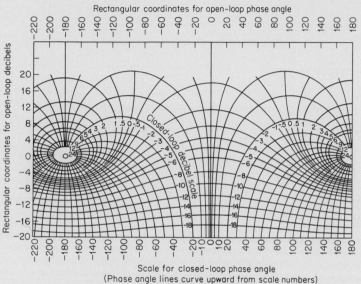

FIG. 55. NICHOLS DIAGRAM

From G. K. Tucker and D. M. Wills, *A Simplified Technique of Control System Engineering* (Philadelphia, Minneapolis-Honeywell Regulator Company, Brown Instruments Division, 1958), p. 187.

feedback loop around this block, the closed-loop equation, again in operator form, is:

$$Z(s)y = N(s)[y_i - y] \tag{5.34}$$

or

$$[Z(s) + N(s)]y = N(s)y_i. \tag{5.35}$$

We see at once that the closed-loop characteristic equation is:

$$Z(s) + N(s) = 0, \tag{5.36}$$

---

* Here s plays three roles: (1) an operator, i.e., a symbol for the operation of differentiation, (2) an algebraic variable in the characteristic equation, and (3) the complex Laplace transform variable. All of these applications have been used before.

where s is again an algebraic quantity. But now let us look at the denominator of the closed-loop transfer function $G(s)/[1 + G(s)]$. Since $G(s) = N(s)/Z(s)$, it is apparent that:

$$1 + G(s) = \frac{Z(s) + N(s)}{Z(s)}. \tag{5.37}$$

It is further apparent that the zeros of $1 + G(s)$ are the roots of the closed-loop characteristic equation and that the poles of $1 + G(s)$ are the roots of the open-loop characteristic equation. Hence the change in phase of the vector $1 + G(s)$ as s moves around a test contour will depend upon the location of the roots of the two characteristic equations.

Suppose then that we had a plot of the open-loop transfer locus $G(s)$ corresponding to one complete counterclockwise journey of s around the closed test contour of Fig. 37. We know that with reference to the point $(-1, j0)$ as origin, this same plot represents the locus of $1 + G(s)$. But we have just seen that the change in phase of the vector $1 + G(s)$ as s moves around the test contour will depend upon the location of the roots of both the closed-loop and open-loop characteristic equations. Specifically, we can say that $G(s)$ will rotate once counterclockwise around the point $(-1, j0)$ in the G-plane for every zero and once clockwise for every pole of $1 + G(s)$ that lies in the right half of the s-plane. Assuming no roots on the imaginary axis, we can thus write:

$$N = z - p, \tag{5.38}$$

where N is the number of counterclockwise rotations of $G(s)$ around the point $(-1, j0)$ in the G-plane, and z is the number of zeros and p the number of poles of $1 + G(s)$ which lie in the right-half of the s-plane. But now we recall that the zeros of $1 + G(s)$ are the roots of the closed-loop characteristic equation, and none of them must be in the right half-plane if the closed-loop system is to be stable. Hence we can say that the stability requirement for the closed-loop system is $z = 0$, and that $G(s)$ must therefore make p clockwise revolutions around the point $(-1, j0)$ in the G-plane as s makes one counterclockwise trip around the test contour. Finally we note that the poles of $1 + G(s)$ are also the poles of $G(s)$ and thus the roots of the open-loop characteristic equation. Assuming the zeros of $N(s)$ are known, p can be determined from the behavior of $G(s)$ around the origin of the G-plane. Then the number of unstable roots of the closed-loop characteristic equation is:

$$z = N + p. \tag{5.39}$$

Again, if pure imaginary roots exist, Eqs. (5.38) and (5.39) do not hold and a modified procedure must be used.

It is apparent that the passage of $G(j\omega_f)$ through the point $(-1, j0)$ in the G-plane corresponds to the passage of $1 + G(j\omega_f)$ through the origin of the $[1 + G(j\omega_f)]$-plane, and thus signifies that the closed-loop characteristic equation has roots on the imaginary axis. This point corresponds to *an open-loop gain of one (or zero db) and a phase lag of 180°*, and is often used in connection with the Bode diagram to define open-loop performance which will be compatible with closed-loop stability. A qualitative statement of this definition is that the open-loop gain must be less than one (or zero db) by the time the phase lag

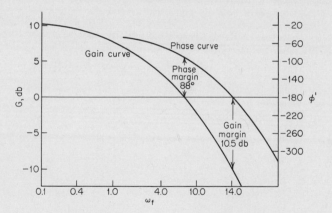

FIG. 56. GAIN AND PHASE MARGINS ON BODE DIAGRAM

reaches 180°. Quantitatively, one can define a gain margin as $-20 \log |G(j\omega_f)|$ at a frequency corresponding to a phase lag ϕ of 180°, and a phase margin as $(180° - \phi)$ at a frequency corresponding to a gain of one (or zero db) (Fig. 56). It is common practice to require open-loop gain margins of 10–20 db and phase margins of 40°–60° in order to safely ensure closed-loop stability.

In terms of the Nyquist transfer locus plot, the requirement that the open-loop gain must be less than one when the phase lag reaches 180° is equivalent to requiring that the locus pass to the right of the critical point $-1 + 0j$ as $j\omega_f$ goes from zero to $+j\infty$. If it passes through this point, then we can expect an undamped oscillatory term in the closed-loop transient response; if it passes to the left of this point, then a growing oscillatory term will appear in the transient response. Another way of stating this stability requirement is that the transfer locus, plotted for $j\omega_f$ from $-j\infty$ to $+j\infty$, shall neither enclose nor pass through the critical point $-1 + 0j$. These simplified stability criteria which are equivalent to those described on page 74 are adequate provided that $G(s)$ has no poles in the right half-plane and this is frequently the case for systems of practical interest.

It is tempting to interpret the critical point $-1 + 0j$ (or open-loop gain of one and phase lag of 180°) in terms of "physical reasoning," particularly for a feedback system in which "recirculation" of the signal is an explicit feature. Suppose, for example, that we introduce a single half-cycle of a sine wave at y_i in the feedback system of Fig. 57, and suppose further that the initial

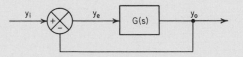

FIG. 57. UNITY GAIN FEEDBACK SYSTEM

conditions are such that y_o will show no transient. Then y_i will appear immediately at y_e, and, if $|G(j\omega_f)| = 1$ and $\phi' = -180°$, it will appear 180° later at y_o (Fig. 58). The error detector will now "invert" y_o, and this inverted signal will appear at y_e. It is now apparent from Fig. 58 that y_e and y_o will continue to oscillate 180° apart without any additional input at y_i. If $|G(j\omega_f)| < 1$, then the signal will be attenuated by a constant fraction on each passage around the loop and the amplitude of oscillation will decay exponentially. On the contrary, if $|G(j\omega_f)| > 1$, then the signal will be amplified on each passage around the loop and the amplitude of oscillation will grow exponentially.

Relationship between frequency response and transient response

If one knows the frequency response of a system, what can be said about its transient response? For first- and second-order systems, the answer is both simple and complete. Here it is obvious that the Bode frequency diagram, interpreted in terms of τ, or of ω_n and ζ, will correspond exactly to a particular transient response of the type shown in Figs. 28 and 32 of Chapter 4. But what about higher-order systems? We have indicated that frequency analysis was of particular value here because of its relative simplicity. The question now arises as to whether this simplicity extends to the prediction of the exact transient response once the frequency response is known. Unfortunately, the answer is that there is no general method of doing this which is both simple and rigorous. An approximate method, often used in engineering applications and based largely upon empirical experience, assumes that the higher-order system

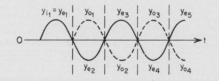

FIG. 58. OSCILLATION IN A FEEDBACK SYSTEM

may be treated as if it were second order. Thus it is assumed that a design based on the frequency method which holds the maximum gain below that of the peak transient overshoot of a second-order system for a particular ζ (Fig. 32) will have a transient response which does not differ significantly from the corresponding second-order transient. This approximation will come close to the truth if the higher-order system has one pair of conjugate complex roots which lie much closer to the imaginary axis than any other roots, for then the transient response will be dominated by this single oscillatory mode and, except for an insignificant initial period, will be that of the corresponding second-order system.

There is, of course, a rigorous method of obtaining the transient response from the frequency response. It involves the evaluation of the inverse Fourier transform integral, but this is not a simple problem. An approximate graphical method for doing this (Floyd's method) is described in texts on control system theory.*

The apparent simplicity of system synthesis by the frequency response method would thus seem to be something of an illusion to the control-system engineer for whom satisfactory transient behavior is generally the most important design criterion. It is therefore not surprising that much recent work in control-system theory has centered about the development of synthesis techniques which permit simultaneous specification of both transient and frequency response. Consideration of such advanced techniques† (e.g., Evans root-locus method, Wiener statistical design technique) is beyond the scope of this work.

Summary

Frequency analysis is concerned with the forced response of linear systems to sinusoidal inputs and with the general question of stability. It is simple to apply and yields results of general significance. It is particularly useful in dealing with higher-order systems where the frequency response can be determined and stability tests applied without the necessity of solving the characteristic equation. It is also useful in synthesis and in determining closed-loop performance and stability from open-loop frequency characteristics. On the other hand, except for first- and second-order systems, there is no simple way to predict the transient response from a knowledge of the frequency response.

* G. S. Brown and D. P. Campbell, *Principles of Servomechanisms* (New York, John Wiley and Sons, Inc., 1948).

† J. G. Truxal, *Automatic Feedback Control System Synthesis* (New York, McGraw-Hill Book Company, Inc., 1955); G. J. Murphy, *Basic Automatic Control Theory* (Princeton, N. J., D. Van Nostrand Company, Inc., 1957).

This chapter completes our consideration of linear, lumped-constant physical systems. We have described their general nature, some of the mathematical tools used to study them, and their responses to certain standard inputs. We have noted that different hardware systems may share a common mathematical model and are in this sense analogous. We have noted the general nature of feedback and have discovered that the form of the differential equation describing a given system does not reveal whether it is a feedback system or not. We have also described various control modes used in man-made regulators to accomplish specific purposes. We now turn to biological systems and will take an introductory look at them in Chapter 6.

CHAPTER 6

Introduction to Biological Control Systems

WHEN the practical demands of World War II required a body of theoretical knowledge to facilitate the design of automatic weapons systems, modern control-system theory was born. Its early development was nearly explosive, for, in the beginning, a considerable stock of nourishing food was readily available. Thus mathematical machinery defining the dynamic behavior of linear electrical and mechanical networks was already highly developed. Once the basic principle of negative feedback was recognized, the synthesis of this knowledge into a form descriptive of closed-loop control systems was readily accomplished. In fact, such a synthesis had already been made for feedback amplifiers by Nyquist and Bode. The linear theory so developed was successfully applied to many practical control problems which arose during and after the war. It is only in recent years, with the development of interest in adaptive control systems, that the limitations of the earlier theory have become increasingly serious.

Such limitations are at once apparent to anyone attempting to describe a biological control system. These systems almost always contain essential nonlinearities arising from individual system components, from parametric feedback loops, or both. In addition, they are frequently characterized by multiple feedback loops arranged in some sort of hierarchy. Complex adaptive behavior is typical of biological systems. Moreover, as pointed out in Chapter 1, the biologist faces an essentially different problem than that encountered by the engineer, for the inductive problem is still very prominent in biology. Relevant variables are often extremely difficult to measure, control, or even identify. System isolation is invariably incomplete so that biological test data are characterized by a relatively high degree of variation, i.e., by a low signal-to-noise ratio. Accordingly, even the steady-state behavior which control engineers regard as trivial is very incompletely understood for many if not most biological systems.

But although some of the details of linear control-system theory may have limited application in biology, the general systems approach remains the same and has much to offer. Not the least of its benefits is the precise identification and rigorous definition of previously vague concepts which formulation of the problem in terms of block diagrams and mathematical "laws" demands. Anyone who has tried this cannot help but be impressed by the insight and clarification which this approach provides.

To a considerable extent, the applicability of simple linear control-system theory to a biological system will depend upon the degree of detail in which we are interested and whether our goal is a special empirical or a general theoretical description. For example, over a restricted amplitude range, the frequency response of a biological "black box" may be indistinguishable from that of a simple linear first- or second-order system. This may be all of the information required for a particular application. On the contrary, it may not satisfy us at all. We may wish to examine the contents of the box in detail and to derive equations for its over-all behavior on theoretical grounds. When we do so we may obtain expressions which are very complex in form but which behave rather simply under certain special conditions, just as relativistic mechanics reduces to Newtonian at "ordinary" velocities. Hence we may approach the analysis of a biological control system in more than one way, and each approach will have its own special advantages and limitations. We shall see how this works out in the particular examples to be considered in the next two chapters.

CHAPTER 7

The Respiratory Chemostat

THE observation that arterial hypercapnia, acidemia, and hypoxemia increase pulmonary ventilation, whereas voluntary hyperventilation produces arterial hypocapnia, alkalemia, and hyperoxemia sets the stage for the concept of a closed-loop respiratory chemostat. If we imagine the system to be a regulator designed to keep arterial pCO_2, (H^+), and pO_2 close to "normal" levels in the face of certain particular disturbances, we can readily describe its operation in qualitative terms. Thus, if a subject breathes 5 percent CO_2 in air, arterial pCO_2 and (H^+) rise. The rise in pCO_2 and (H^+) stimulates ventilation. The increase in ventilation in turn lowers arterial pCO_2 and (H^+) back toward normal. If we inject a strong acid into the blood of an animal, arterial (H^+) rises. The rise in (H^+) stimulates ventilation. The increase in ventilation decreases arterial pCO_2 below normal and thus lowers (H^+) back toward normal. Finally, if a subject breathes an oxygen-deficient mixture, arterial pO_2 falls. This fall in pO_2 stimulates ventilation. The increase in ventilation raises arterial pO_2 back toward normal. In so doing, it also produces hypocapnia and alkalemia.

It is thus evident that we are dealing with a feedback regulator and that this regulator is concerned not only with one but with at least three controlled quantities. It is also evident that the system shows steady-state error in the presence of the disturbances described above. In some instances, e.g., arterial anoxemia, the response designed to correct the primary error will lead to undesirable errors in other controlled quantities. Let us now attempt to draw a block diagram for this system.

Block diagram

As a modest beginning, consider the diagram in Fig. 59. In it, we have preserved the general pattern of Fig. 13 and have recognized two blocks representing, respectively, a controlling and a controlled system. The former receives three command and three feedback signals and generates a single output which

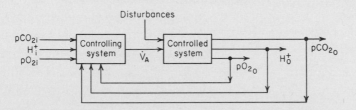

FIG. 59. BLOCK DIAGRAM OF RESPIRATORY CHEMOSTAT

we take as alveolar ventilation $\dot{V}_a$. This controlling signal (or manipulated variable) provides the input to the controlled system which generates the three outputs (or controlled variables): arterial pCO_2, (H^+), and pO_2. The latter in turn are fed back to the controlling system. The "disturbances" shown entering the controlled system block are those we have considered above, i.e., CO_2 excess or oxygen deficiency in inspired gas, and introduction of a strong acid into the blood.

It is at once obvious to any physiologist that this diagram conceals many details anonymously lumped within the two component blocks. Thus the chemical signals shown entering the controlling system actually operate on neural receptors located at various anatomical sites. These receptors generate neural signals which enter the respiratory center via a variety of paths. Depending upon these signals as well as upon neural afferents from the lungs, the center sends out periodic motor signals to the respiratory muscles which in turn generate a forcing pressure to drive the respiratory pump. Depending upon the pump impedance, a certain total pulmonary ventilation will result. Depending upon the behavior of the "dead air," this will correspond finally to a certain alveolar ventilation $\dot{V}_a$.

Thus the contents of the controlling system block are quite complex and evidently consist of a number of subsystems each with its own input, output, and properties. In like fashion, many details have been lumped within the controlled system block. Included among them are the pulmonary diffusion process, the transport of gases by the blood, including the chemistry of the blood buffer systems, and gas exchange in the tissues. No one will deny that a complete description of the respiratory regulator requires a detailed analysis of all of these components. However, it is possible to understand many important features of over-all system behavior without having to do this. This is the approach we shall use in what follows.

The steady-state chemostat

Let us begin by considering the operation of the chemostat only in the steady state. We would like to obtain a law or transfer function for each block defining the dependence of output(s) upon input(s), and then to combine them to obtain the behavior of the closed-loop system. Because we shall ignore the details of the respiratory cycle and examine only steady-state responses to constant forcings, these laws will take the form of algebraic equations.

It is perhaps more than coincidental that the first mathematical model of the respiratory chemostat was, like control-system theory in general, a child of World War II. Consideration of such practical questions as the oxygen requirements of pilots at high altitude, the possible use of CO_2 to counteract anoxia, etc., led Gray in 1945 to the formulation of his "multiple factor theory."[*] Although neither the term feedback nor the block-diagram representation was employed, this theory did in fact treat the respiratory regulator as a closed-loop system.

To obtain the law of the controlling system (which he called the chemical ventilation equation), Gray assumed that each of the three arterial chemical agents, pCO_2, (H^+), and pO_2, ex-

[*] J. S. Gray, "The multiple factor theory of respiratory regulation," AAF School of Aviation Medicine, Randolph Field, Texas, Project No. 386, Report No. 1, May 7, 1945.

erted an independent effect on $\dot{V}_a$, and that these effects were additive. By using empirical data on the response to CO_2 inhalation [during which pCO_2 and (H^+) move in the same direction] together with data on metabolic disturbances in acid-base balance (during which pCO_2 and (H^+) move in opposite directions), he was able to separate the partial effects of pCO_2 and (H^+) and assign a gain coefficient to each. Then, using data on arterial anoxemia and allowing for the inhibitory effect of the respiratory alkalosis present, he derived an expression for the partial effect of pO_2. The steady-state law so obtained for the controlling system was:

$$\dot{V}_a = 1.1 \; (H^+) + 1.31 \; pCO_2 - 90 + 10.6 \times 10^{-8}(104 - pO_2)^{4.9},$$
$$(7.1)$$

where $\dot{V}_a$ is alveolar ventilation in liters/min., (H^+) is arterial hydrogen ion concentration in $m\mu M$/liter, and pCO_2 and pO_2 are arterial gas tensions in mm Hg.

Although new evidence derived from combinations of CO_2 inhalation and anoxemia indicates that the pCO_2 and pO_2 effects are not independent and that there may be a threshold for pCO_2, Eq. (7.1) will suffice for present purposes. Note that it has absolutely nothing to say about many details which might well be considered important in any general consideration of the physiology of respiration. No attempt has been made to analyze the contents of the controlling system block in detail. We merely say that if arterial (H^+), pCO_2, and pO_2 have certain values, then $\dot{V}_a$ is determined. Implicit in this statement is the assumption that the box contents are normal.

Let us now consider the controlled system. Here we need expressions defining the dependence of arterial pCO_2, (H^+), and pO_2 upon alveolar ventilation, composition of inspired gas, and buffer properties of blood. It turns out to be relatively easy to obtain theoretical expressions for alveolar pCO_2 and pO_2 based upon the principle of continuity. They are as follows:

$$pCO_2 = pCO'_2 + \frac{KMR}{\dot{V}_a} \tag{7.2}$$

$$pO_2 = pO'_2 - \frac{KMR}{\dot{V}_a}, \tag{7.3}$$

where pCO'_2 and pO'_2 are "tracheal" CO_2 and O_2 tensions, respectively, MR is metabolic gas exchange rate which is assumed equal for CO_2 and O_2 (i.e., $RQ = 1$), and K includes barometric pressure along with a constant to establish compatible volume units.

Equations (7.2) and (7.3) convey a very simple "dilution" message. They say that when a given volume of tracheal air of CO_2 tension pCO'_2 and O_2 tension pO'_2 is changed to an equal volume $\dot{V}_a$ of alveolar air by having MR cc of oxygen replaced by MR cc of CO_2, its CO_2 tension increases by $KMR/\dot{V}_a$ and its oxygen tension decreases by an equal amount. If we now assume that alveolar and arterial gas tensions are equal, we have two of our three required expressions. We still need a relationship between (H^+) and $\dot{V}_a$. We can obtain it indirectly by examining the relationship between (H^+) and pCO_2 determined by the blood buffer systems. A rigorous theoretical analysis of this multiphase, multicomponent physicochemical equilibrium is extremely complex and yields very formidable equations. However, for many practical purposes, we can employ a linear approximation of the following form:

$$(H^+) = apCO_2 + b, \tag{7.4}$$

where a and b are variable parameters which depend upon the standard bicarbonate content, oxygen capacity, and oxygen saturation of the blood. It thus becomes possible to express (H^+) as a function of pCO_2 and pO_2 and thus of $\dot{V}_a$. Equations (7.2)–(7.4) are the equations of the controlled system. If they are solved simultaneously with the controller equation (7.1), closed-loop relationships defining the steady-state responses of $\dot{V}_a$, pCO_2, (H^+), and pO_2 to CO_2 inhalation, metabolic disturbances in acid-base balance, or to low pO'_2 can be obtained. This is what Gray did.

Despite the omission of many details, this mathematical model for the steady-state operation of the "respiratory chemostat" gives very important information about and new insights into the regulation of breathing. An understanding of its operation is basic to the investigation of other respiratory responses including the still mysterious hyperpnea of exercise.

The reception which Gray's theory received from respiratory physiologists in 1945–46 should perhaps have been expected. Few welcomed it as a valuable contribution, but many categorically denied the value of any such mathematical analysis in biology. Some demanded "more dogs and less talk," while others implied that there was nothing here that everyone didn't know already. The fact is that this theory has directly or indirectly stimulated much of the productive work in respiratory regulation which has been done over the past 15 years. John Pierce of the Bell Telephone Laboratories describes such a situation nicely in his recent book *Symbols, Signals, and Noise.* Speaking of the origin of scientific ideas, he points out that when someone else actually solves a problem that we merely had ideas about, we think that we understood it all along, for we genuinely believe that because we had stated many of the required concepts in juxtaposition, we must really have reached the general conclusion. Yet it is true that formulations which may seem obvious when stated have waited years for the insight that enabled someone to make the statement.

Relation to linear feedback theory. Let us now pause and inquire how this steady-state chemostat compares with the simple physical system illustrated in Fig. 13. Limiting our consideration of the latter also to steady states, we can draw the block diagram

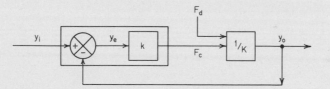

FIG. 60. ZERO-ORDER LINEAR FEEDBACK SYSTEM

of Fig. 60. Perhaps the most obvious difference between Figs. 59 and 60 is the presence of three loops in the former and only one in the latter. Let us eliminate this difference at once by restricting our consideration of the chemostat to only one type of disturbance, i.e., to CO_2 inhalation. Then we can neglect the pO_2 loop and incorporate the (H^+) loop into the pCO_2 loop via

Eq. (7.4) to obtain the single-loop chemostat of Fig. 61. If we ignore the box contents, Figs. 60 and 61 obviously conform to a single pattern.

Let us now compare the box contents starting with the controlling system. The error detector shown in the physical system

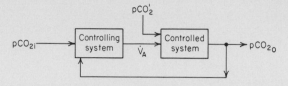

FIG. 61. SINGLE-LOOP RESPIRATORY CHEMOSTAT

is an actual tangible device which the engineer builds into his controller. If the system is designed to be a regulator rather than a servo mechanism, the set point y_i will be chosen by adjusting a knob on the controller. If this set point is not equal to zero, then, as noted in Chapter 2, we must add a reference value to the controller output to zero the steady-state error when $F_d = 0$. In the present case, this value is given by $F_R = Ky_i$. The corresponding block diagram is shown in Fig. 62. Again,

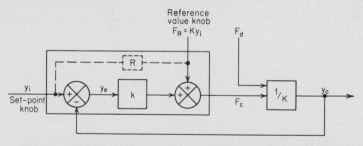

FIG. 62. REGULATOR WITH SET-POINT AND REFERENCE VALUE ADJUSTMENTS

as noted in Chapter 2, we may have separate adjustment knobs for set point and reference value, or a single knob may serve both purposes. Now let us ask whether any of these items which characterize man-made controllers can be identified in the controlling system of the respiratory chemostat.

Obviously we cannot point out anatomical entities which

serve as set-point knob, error detector, and reference-value knob. Also, we have no way of knowing what the set point really is. But if the concepts of error detector, set point, and reference values are convenient ones, then there is no reason why we should not use them provided we recognize their arbitrariness. For example, based on experimental data during CO_2 inhalation, we might write the controller equation of our single-loop chemostat as follows:

$$\dot{V}_a = 2p CO_{2o} - 75. \tag{7.5}$$

This equation explicitly recognizes neither a set point, error detector, nor reference value. Yet they are all implicitly present as the following manipulation shows:

$$\dot{V}_a = 2(p CO_{2o} - 40) + 5. \tag{7.6}$$

Here we have chosen $p CO_{2i} = 40$ as our set point corresponding to the normal value when $p CO'_2 = 0$. The reference value is then defined from Eq. (7.2) as $\dot{V}_{a(r)} = KMR/p CO_{2i} = 5$. This choice is arbitrary but practical. Actually we have no way of knowing whether $p CO_{2i}$ is actually 40. For example, we could equally well choose $p CO_{2i} = 50$ and write (7.5) as:

$$\dot{V}_a = 2(p CO_{2o} - 50) + 25 \tag{7.7}$$

and this corresponds to the assumption that the system was designed to show zero error in the presence of $p CO'_2 = 42$.

Having written (7.5) to include an explicit error detector, we at once notice a difference between this error detector and that illustrated in Fig. 62. It is apparent that the $p CO_2$ error in (7.6) or (7.7) is the negative of the y error in Fig. 62. This is necessary because, as implied by Eq. (7.2), a sign (directional) change in the signal occurs during its passage through the controlled system of the chemostat, i.e., $p CO_{2o}$ decreases when $\dot{V}_a$ increases. Unfortunately, it turns out that this sign-changing operation is accomplished in such a way as to destroy the simplicity of our chemostat.

First we note that if it had been accomplished in linear fashion, the simplicity would have been preserved. If our controlled-system equation were:

$$p CO_{2o} = (A - B\dot{V}_a) + p CO'_2 \tag{7.8}$$

and the controller equation:

$$\dot{V}_a = k(pCO_{2o} - pCO_{2i}) + \dot{V}_{a(r)}, \qquad (7.9)$$

where $\dot{V}_{a(r)}$ is defined as $(A - pCO_{2i}/B)$, then the closed-loop equation is easily shown to be:

$$pCO_{2o} = pCO_{2i} + \frac{1}{1 + Bk} pCO'_2 \qquad (7.10)$$

or $\qquad (pCO_{2o} - pCO_{2i}) = pCO_{2e} = \dfrac{1}{1 + Bk} pCO'_2.$ $\qquad (7.11)$

Just as in the physical system of Fig. 62, the steady-state error is zero when $pCO'_2 = 0$, and the closed-loop error transfer function for regulator operation is a familiar one:

$$\frac{pCO_{2(e)}}{pCO'_2} = \frac{1}{1 + Bk}. \qquad (7.12)$$

But let us now return to reality! We already know that the change in signal direction produced by the controlled system is accomplished through the nonlinear (hyperbolic) operation of Eq. (7.2):

$$pCO_{2(o)} = \frac{KMR}{\dot{V}_a} + pCO'_2. \qquad (7.2)$$

What does this do to the transfer-function concept which provided such a simple and convenient description of linear systems? First we note that it is easy to obtain transfer functions for the controller equation (7.6) and for the hypothetical linear controlled-system equation (7.8). The fact that these two equations are not direct proportions does not destroy their basic simplicity but merely requires a scale-shifting operation.

Thus we can represent (7.6) in the following transfer-function form:

$$\frac{\dot{V}_a - 5}{pCO_{2(o)} - 40} = 2 \qquad (7.13)$$

corresponding to the block diagram of Fig. 63. In like fashion, Eq. (7.8) becomes:

$$\frac{pCO_{2o} - (pCO'_2 + A)}{\dot{V}_a} = -B \qquad (7.14)$$

and its block diagram appears in Fig. 64. In each case, the transfer function of the block is a *constant* which when multiplied by the input yields the output.

The scale-shifting operations are represented by simple summing points. But what happens if we attempt to put (7.2) into

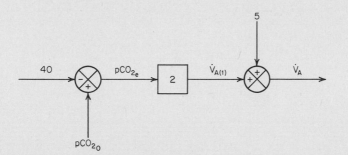

FIG. 63. BLOCK DIAGRAM OF RESPIRATORY CONTROLLING SYSTEM

transfer-function form? Subtracting pCO'_2 from both sides and dividing through by $\dot{V}_a$ yields:

$$\frac{pCO_{2o} - pCO'_2}{\dot{V}_a} = \frac{KMR}{\dot{V}_a{}^2} \qquad (7.15)$$

and we at once see an essential difference. The transfer function

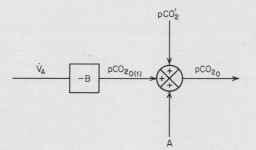

FIG. 64. BLOCK DIAGRAM OF HYPOTHETICAL LINEAR
RESPIRATORY CONTROLLED SYSTEM

defined by (7.15) is not a constant at all, *but is a function of the input $\dot{V}_a$.* This provides us with a definition for a nonlinear block, i.e., it is a block whose transfer function is dependent upon the input.

If we now combine (7.2) with controller equation (7.6) taking pCO_{2i} as set point, k as controller gain, and KMR/pCO_{2i} as reference value, the beautiful formal simplicity of the linear system disappears. The closed-loop equation is a quadratic whose solution by quadratic formula is:

$$pCO_{2o} = \frac{pCO_{2i} + pCO'_2 - (KMR/kpCO_{2i})}{2}$$

$$+ \left\{ \frac{[pCO_{2i} + pCO'_2 - (KMR/kpCO_{2i})]^2}{4} \right.$$

$$\left. - [pCO_{2i}pCO'_2 - (KMRpCO'_2/kpCO_{2i}) - (KMR/k)] \right\}^{1/2}$$

$$(7.16)$$

This is a far cry from the simple relationships of Eqs. (2.33) and (7.12). No longer is it possible to characterize the closed-loop behavior by a simple transfer function. We should nevertheless point out that despite the considerable formal difference between Eqs. (7.16) and (7.12), the behavior of the two may not differ greatly over restricted ranges. Thus the difference in steady-state errors of the two systems may remain within 2 mm Hg over a pCO'_2 range from 0 to 40 mm Hg (Fig. 65).

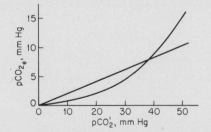

FIG. 65. STEADY-STATE ERROR AS FUNCTION OF pCO'_2 FOR HYPOTHETICAL LINEAR AND ACTUAL NONLINEAR RESPIRATORY CHEMOSTAT

The formal inconveniences provided by the nonlinearity of the steady-state chemostat hint of even greater difficulties to come if we should attempt a dynamic analysis of this closed-loop regulator. The remainder of this chapter will be devoted to such an analysis. There we shall find that although the isolated

controlled system when forced by CO_2 in the inspired air behaves like the linear second-order systems of Chapter 4, the
closed-loop equations are nonlinear because of "feedback through
parameters." Much of the mathematical simplicity is thereby
destroyed, and an increasingly important role must be assigned
to computers in studying the behavior of such systems.

The dynamic chemostat*

Transient response to CO_2 *inhalation.* In the steady-state chemostat, it was convenient to assume that the controlling system
derived its input signals from the levels of pCO_2, (H^+), and pO_2
in arterial blood. This did not imply that the receptors were
actually located there, but only that in the particular steady-
state situations considered, the arterial levels correlated closely
with the effective levels at the true receptor sites wherever these
might be. It was clearly recognized that this correlation could
not be expected to hold under all conditions. One in which it
does not hold is the transient response to CO_2 inhalation, and
the first dynamic model of the respiratory chemostat was formulated to account for this behavior.

Thus, if we look at the time course of arterial pCO_2 and
pulmonary ventilation following a step-function change in inspired CO_2 concentration (Fig. 66), we see at once that we have
a new problem. Clearly, the single-valued relationship between
ventilation and arterial pCO_2 which holds in the steady state
[Eq. (7.6)] does not hold in the transient. One possible explanation for this might be that the controlling system actually receives its input signal from the levels of pCO_2 and (H^+) in the
tissues of the respiratory center, and that changes in these
levels lag behind those in arterial blood because of the large
tissue storage capacity for CO_2. Let us construct a model based
on this assumption.

To simplify the analysis, we shall make a number of other
assumptions as well. These are: (1) that the lungs can be re-

* F. S. Grodins, J. S. Gray, K. R. Schroeder, A. L. Norins, and R. W. Jones,
"Respiratory responses to CO_2 inhalation: a theoretical study of a nonlinear biological regulator," *J. Appl. Physiol.* 7 (1954): 283.

garded as a box of constant volume, zero dead space, and homogeneous content ventilated by a continuous stream of gas; (2) that the RQ is one at every instant; (3) that blood transport lags are negligible; (4) that the respiratory center can be lumped with all other tissues into a single homogeneous reservoir having

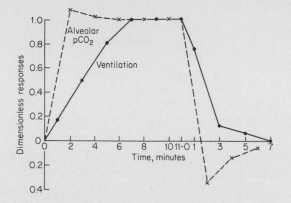

FIG. 66. RESPONSE OF VENTILATION AND ALVEOLAR $p CO_2$ TO
INHALATION OF 5.43 PERCENT CO_2

Data of Padget. From F. S. Grodins, J. S. Gray, K. R. Schroeder, A. L. Norins, and R. W. Jones, "Respiratory responses to CO_2 inhalation: a theoretical study of a nonlinear biological regulator," *J. Appl. Physiol*. 7 (1954): 283.

a constant blood flow; (5) that arterial blood, venous blood, and "tissue" have the same linearized CO_2 absorption curve; (6) that expired air, alveolar air, and arterial blood are in continuous CO_2 equilibrium as are "tissue" and venous blood; and (7) that the controlling system is a simple proportional controller containing no dynamic elements. Let us begin our analysis with the controlled system.

The controlled system. A hardware diagram of this system incorporating our first six assumptions is shown in Fig. 67. It consists of two constant-volume reservoirs connected by the circulating blood. Each reservoir can exchange gases with the blood through a diffusion membrane. Carbon dioxide is poured into the tissue reservoir at a rate corresponding to metabolic CO_2 production. Gas of arbitrary composition is pumped into the lung reservoir at a rate corresponding to alveolar ventilation.

It is not difficult to write a set of equations defining the dynamic behavior of this system. To do so, we use the same general principles of equilibrium and continuity which provided the basis for Newton's laws of motion and Kirchhoff's laws of electrical networks in Chapter 2. Beginning with the lung or alveolar reservoir, we write a "CO_2 continuity" equation which

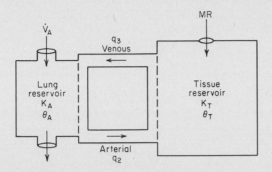

FIG. 67. LUNG, BLOOD, TISSUE CO_2 EXCHANGE SYSTEM

From F. S. Grodins, J. S. Gray, K. R. Schroeder, A. L. Norins, and R. W. Jones, "Respiratory responses to CO_2 inhalation: a theoretical study of a nonlinear biological regulator," *J. Appl. Physiol.* 7 (1954): 283.

states that the rate of change of alveolar CO_2 concentration $\dot{\theta}_A$ is equal to the difference between the rates of CO_2 inflow and outflow by all paths divided by the volume of the reservoir K_A:

$$\dot{\theta}_A = \frac{1}{K_A} [\dot{V}_A F_{CO_2}{}^I + q_3 - q_1 - q_2]. \tag{7.17}$$

Carbon dioxide enters the lung via the inspired gas at a rate equal to the product of inspired fraction $F_{CO_2}{}^I$ and ventilation $\dot{V}_A$ and via venous (pulmonary arterial) blood at a rate q_3. It leaves the lung via expired air at rate q_1 and arterial (pulmonary venous) blood at rate q_2. We next write another continuity equation for the tissue reservoir: the rate of change of tissue CO_2 concentration $\dot{\theta}_T$ is equal to the rate of metabolic CO_2 production MR plus the rate of entry of CO_2 in arterial blood q_2 minus its rate of exit in venous blood q_3 all divided by the volume of the reservoir K_T:

$$\dot{\theta}_T = \frac{1}{K_T} [MR + q_2 - q_3]. \tag{7.18}$$

Finally, we write three equilibrium relations. The first expresses the equality of alveolar and expired CO_2 concentrations:

$$\theta_A = \frac{q_1}{\dot{V}_A}. \tag{7.19}$$

The second expresses the equilibrium between arterial and alveolar concentrations via the linearized CO_2 absorption curve:

$$\frac{q_2}{Q} = BA_s(\theta_A) + A_i, \tag{7.20}$$

where Q is cardiac output, B is barometric pressure, and A_s and A_i are the slope and intercept, respectively, of the linear absorption curve. The third expresses the equality of tissue and venous CO_2 concentrations:

$$\theta_T = \frac{q_3}{Q}. \tag{7.21}$$

We have thus obtained a set of five simultaneous equations in the five unknowns: θ_T, θ_A, q_1, q_2, and q_3. Only two of the equations, (7.17) and (7.18), are differential equations, the other three being algebraic. We now wish to solve this set in order to define the behavior of the isolated controlled system. This can be done in various ways, but in any case we must first decide which of the five dependent variables shall be regarded as the system output (s), and which of the nine independent variables ($\dot{V}_A$, $F_{CO_2}{}^I$, MR, K_A, K_T, Q, B, A_s, A_i) shall be regarded as the system forcing (s). Let us begin by choosing θ_T and θ_A as our system outputs, and $F_{CO_2}{}^I(t)$ as our system forcing, thus implying that $\dot{V}_A$, MR, K_A, K_T, Q, B, A_s, and A_i are time invariant constants. We can then combine our five basic equations to obtain a differential equation in the single dependent variable θ_T or θ_A. Thus, if we algebraically solve (7.19) for q_1, (7.20) for q_2, (7.21) for q_3, and substitute these values in (7.17) and (7.18), we obtain from (7.17):

$$\dot{\theta}_A = \frac{1}{K_A} [\dot{V}_A(F_{CO_2}{}^I - \theta_A) + Q(\theta_T - BA_s\theta_A - A_i)] \tag{7.22}$$

and from (7.18):

$$\dot{\theta}_T = \frac{1}{K_T} [MR - Q(\theta_T - BA_s\theta_A - A_i)]. \tag{7.23}$$

We now solve (7.23) for θ_A to get:

$$\theta_A = \frac{1}{QBA_s} (K_T \dot{\theta}_T - MR - QA_i + Q\theta_T) \qquad (7.24)$$

and differentiate (7.24) to obtain:

$$\dot{\theta}_A = \frac{1}{QBA_s} (K_T \ddot{\theta}_T + Q\dot{\theta}_T). \qquad (7.25)$$

Then, substituting (7.24) and (7.25) into (7.22), combining terms and rearranging to "standard form," we obtain the desired equation for θ_T:

$$\frac{K_A K_T}{Q \dot{V}_A} \ddot{\theta}_T + \left[\frac{K_A}{\dot{V}_A} + \frac{K_T BA_s}{\dot{V}_A} + \frac{K_T}{Q} \right] \dot{\theta}_T + \theta_T$$
$$= BA_s F_{CO_2}{}^I(t) + \frac{BA_s MR}{\dot{V}_A} + \frac{MR}{Q} + A_i. \qquad (7.26)$$

Having obtained an equation for θ_T, there are various possibilities for finding θ_A. By far the easiest is to solve (7.26) for θ_T and then use Eq. (7.24) to obtain θ_A directly in terms of θ_T and $\dot{\theta}_T$. However, if we wish, we can obtain an equation for θ_A by methods analogous to those used to get (7.26) for θ_T. In this case we solve (7.22) for θ_T, differentiate the result to get $\dot{\theta}_T$, and substitute these expressions in (7.23). The result is:

$$\frac{K_A K_T}{Q \dot{V}_A} \ddot{\theta}_A + \left[\frac{K_A}{\dot{V}_A} + \frac{K_T BA_s}{\dot{V}_A} + \frac{K_T}{Q} \right] \dot{\theta}_A + \theta_A$$
$$= F_{CO_2}{}^I(t) + \frac{K_T}{Q} \dot{F}_{CO_2}{}^I(t) + \frac{MR}{\dot{V}_A}. \qquad (7.27)$$

Equations (7.26) and (7.27) are the equations of the isolated controlled system under the assumption that the forcing term is $F_{CO_2}{}^I(t)$. Let us now solve these equations to obtain $\theta_T(t)$ and $\theta_A(t)$ when $F_{CO_2}{}^I(t)$ is a step function of magnitude $(F_{CO_2}{}^I)_1$.

First we note that both equations are second-order linear differential equations with constant coefficients, i.e., exactly the same type that we encountered in dealing with physical systems in earlier chapters. Hence we should be able to solve them by either the classical or Laplace transform method and also to identify an analogous electrical or mechanical system.

Recalling first the classical method, we know that the solution will consist of a complementary function or transient response, and a particular integral or forced response. We also know that the former is the general solution of the corresponding homogeneous equation, i.e., the equation obtained by replacing the right side by zero. It is at once apparent that both (7.26) and (7.27) have the same homogeneous equation, namely,

$$\frac{K_A K_T}{Q \dot{V}_A} \ddot{\theta} + \left[\frac{K_A}{\dot{V}_A} + \frac{K_T B A_s}{\dot{V}_A} + \frac{K_T}{Q} \right] \dot{\theta} + \theta = 0. \tag{7.28}$$

Let us now recall from Chapter 4 that the nature of the solution of (7.28) is critically dependent upon the value of the damping ratio ζ. The definition of this dimensionless parameter given in Chapter 2 in connection with a particular mechanical system can now be generalized to apply to any second-order system. Thus, if the general form of (7.28) is:

$$K_2 \ddot{\theta} + K_1 \dot{\theta} + \theta = 0, \tag{7.29}$$

then ζ is defined as:

$$\zeta \equiv \frac{K_1}{2(K_2)^{1/2}}. \tag{7.30}$$

Applying (7.30) to (7.28), we can obtain the damping ratio of the lung-blood-tissue CO_2 exchanger. It is:

$$\zeta = \frac{Q K_A + Q K_T B A_s + \dot{V}_A K_T}{2(\dot{V}_A Q K_A K_T)^{1/2}}. \tag{7.31}$$

If we now substitute appropriate values (Table 5) for the con-

TABLE 5

	Normal	Double $\dot{V}_A$	Double Q	Halve K_T	Halve K_A
Tissue volume, K_T (liters)	40	40	40	20	40
Cardiac output, Q (liters/min.)	6	6	12	6	6
Lung volume, K_A (liters)	3	3	3	3	1.5
Ventilation, $\dot{V}_A$ (liters/min.)	5	10	5	5	5
Barometric pressure, B (mm Hg)	760	—	—	—	—
Absorption curve slope, A_s (liters/liter/mm Hg)	0.00425	—	—	—	—
Damping ratio, ζ	8.28	7.04	10.5	6.0	11.6
T_1 (min.)	0.12	0.10	0.07	0.12	0.06
T_2 (min.)	32.94	19.79	29.70	16.74	32.75

stants in Eq. (7.31), we find that our system is greatly over-damped, i.e., ζ is at least 6.0, so that the system is very stable.

It turns out to be most convenient to describe such over-damped systems not in terms of ζ and ω_n, but rather in terms of two effective time constants, T_1 and T_2. We first note that we can write (7.28) in terms of three time constants as follows:

$$(\tau_1\tau_2)\ddot{\theta} + (\tau_1 + \tau'_{12} + \tau_2)\dot{\theta} + \theta = 0, \qquad (7.32)$$

where

$$\tau_1 \equiv \frac{K_A}{\dot{V}_A}, \quad \tau_2 \equiv \frac{K_T}{Q}, \quad \text{and} \quad \tau'_{12} \equiv \frac{K_T}{\dot{V}_A}\,(BA_s).$$

Then, if we define the effective time constants T_1 and T_2 in such a way that:

$$T_1T_2 = \tau_1\tau_2 \qquad (7.33)$$

and

$$T_1 + T_2 = \tau_1 + \tau'_{12} + \tau_2, \qquad (7.34)$$

we can finally write (7.28) as:

$$(T_1T_2)\ddot{\theta} + (T_1 + T_2)\dot{\theta} + \theta = 0. \qquad (7.35)$$

We recall from Chapter 3 that the general solution of (7.35) is

$$\theta = C_1 e^{r_1 t} + C_2 e^{r_2 t}, \qquad (7.36)$$

where r_1 and r_2 are the roots of the algebraic characteristic equation corresponding to (7.35), i.e., the roots of:

$$(T_1T_2)r^2 + (T_1 + T_2)r + 1 = 0. \qquad (7.37)$$

Equation (7.37) is obviously factorable:

$$(T_1r + 1)(T_2r + 1) = 0 \qquad (7.38)$$

and the roots are:

$$r_1 = -1/T_1 \qquad (7.39)$$
$$r_2 = -1/T_2. \qquad (7.40)$$

Hence the complementary function or transient response θ_t for both (7.26) and (7.27) is:

$$\theta_t = C_1 e^{-t/T_1} + C_2 e^{-t/T_2}. \qquad (7.41)$$

The next step in the classical solution is to find the particular integral or forced response θ_p. Since the right side of (7.26) is a constant, the methods of Chapters 3 and 4 at once tell us that:

$$\theta_{T(P)} = BA_s(F_{CO_2}{}^I)_1 + \frac{BA_sMR}{\dot{V}_A} + \frac{MR}{Q} + A_i \qquad (7.42)$$

and this, of course, is the steady-state value $\theta_{T(ss)}$. It will now pay us to pause a moment and take a closer look at the right side of (7.26), for we have not really treated this particular situation explicitly before. Although it is true that $\theta_{T(ss)}$ is a constant for $t > 0$, part of it (i.e., the last three terms) has the same value for all times including $t < 0$, and so really represents an initial condition on θ_T rather than a step forcing. We can handle such a situation in either of two ways: (1) include these constant terms in the step forcing as implied by (7.42) and also as initial conditions on θ_T, or (2) omit them from both and simply add them on to the solution at the end. Both methods reveal that such constant terms merely indicate a scale-shifting operation.

Using the first method, we write the general solution of (7.26) as:

$$\theta_T = C_1 e^{-t/T_1} + C_2 e^{-t/T_2} + \theta_{T(ss)}, \tag{7.43}$$

where $\theta_{T(ss)}$ is defined by (7.42) and C_1 and C_2 are to be evaluated by taking $\theta_T = [(BA_s MR/\dot{V}_A) + (MR/Q) + A_i] \equiv \theta_{T_0}$, and $\dot{\theta}_T = 0$ at $t = 0$. The methods of Chapter 4 yield the following values for C_1 and C_2:

$$C_1 = [\theta_{T_0} - \theta_{T(ss)}] \left(\frac{T_1}{T_1 - T_2} \right) \tag{7.44}$$

$$C_2 = [\theta_{T_0} - \theta_{T(ss)}] \left(\frac{-T_2}{T_1 - T_2} \right) \tag{7.45}$$

and our final equation is:

$$\theta_T = [\theta_{T_0} - \theta_{T(ss)}] \left[\frac{T_1}{T_1 - T_2} e^{-t/T_1} - \frac{T_2}{T_1 - T_2} e^{-t/T_2} \right] + \theta_{T(ss)} \tag{7.46}$$

or, as a dimensionless ratio,

$$\frac{\theta_{T(ss)} - \theta_T}{\theta_{T(ss)} - \theta_{T_0}} = \frac{T_1}{T_1 - T_2} e^{-t/T_1} - \frac{T_2}{T_1 - T_2} e^{-t/T_2}. \tag{7.47}$$

The fact that (7.47) gives exactly the same curve for any values of θ_{T_0} and $\theta_{T(ss)}$ illustrates why engineers concerned with dynamics are indifferent to the steady-state values at either end of the transient response. For example, they would not hesitate to neglect the last three terms on the right side of (7.26) and take $\theta_{T_0} = 0$, despite the horrible implications that this might have for a physiologist (i.e., $MR = 0$)!

What about the solution of (7.27) for θ_A? As we have already noted, the complementary function is given by (7.41). However, we encounter something new when we seek a particular integral, for not only $F_{CO_2}{}^I(t)$ but also its derivative $\dot{F}_{CO_2}{}^I(t)$ appears on the right side of (7.27). Now as noted in Chapter 3, if $F_{CO_2}{}^I(t)$ is a step function, then its derivative $\dot{F}_{CO_2}{}^I(t)$ is an impulse or delta function. It turns out that the effect of an impulse of magnitude I can be accounted for by adding the quantity I/K_n to the initial condition on the $(n - 1)$th derivative, where K_n is the coefficient of the highest derivative with the equation written in standard form as in (7.29). Hence, to obtain our particular integral, we can regard the right side of (7.27) as a constant and write:

$$\theta_{A(P)} = (F_{CO_2}{}^I)_1 + \frac{MR}{\dot{V}_A} \equiv \theta_{A(ss)}. \qquad (7.48)$$

Incidentally, if we multiply this steady-state solution by barometric pressure, it becomes identical with Eq. (7.2) previously derived for the steady-state chemostat. Now the general solution of (7.27) can be written as:

$$\theta_A = C_3{}^{-t/T_2} + C_4 e^{-t/T_2} + \theta_{A(ss)} \qquad (7.49)$$

and it only remains to evaluate C_3 and C_4 from our initial conditions.

When $t = 0$, $\theta_A = MR/\dot{V}_A \equiv \theta_{A_0}$, and $\dot{\theta}_A$ will have a value dictated by the impulse function $\tau_2(\dot{F}_{CO_2}{}^I)_1(t)$ as noted above. The magnitude of this impulse is evidently $\tau_2[\theta_{A(ss)} - \theta_{A_0}]$, and since $K_n = T_1T_2$, $\dot{\theta}_{A_0} = (\tau_2/T_1T_2)[\theta_{A(ss)} - \theta_0]$. Using these initial conditions, we evaluate C_3 and C_4 as:

$$C_3 = [\theta_{A_0} - \theta_{A(ss)}]\left(\frac{T_1 - \tau_2}{T_1 - T_2}\right) \qquad (7.50)$$

$$C_4 = -[\theta_{A_0} - \theta_{A(ss)}]\left(\frac{T_2 - \tau_2}{T_1 - T_2}\right) \qquad (7.51)$$

and our final equation becomes:

$$\theta_A = [\theta_{A_0} - \theta_{A(ss)}]\left[\frac{T_1 - \tau_2}{T_1 - T_2} e^{-t/T_1} - \frac{T_2 - \tau_2}{T_1 - T_2} e^{-t/T_2}\right] + \theta_{A(ss)}$$

(7.52)

or, as a dimensionless ratio:

$$\frac{\theta_{A(ss)} - \theta_A}{\theta_{A(ss)} - \theta_{A_0}} = \frac{T_1 - \tau_2}{T_1 - T_2} e^{-t/T_1} - \frac{T_2 - \tau_2}{T_1 - T_2} e^{-t/T_2}. \quad (7.53)$$

It is perhaps illuminating to look at (7.52) from another point of view by writing it as follows:

$$\theta_A = \left\{[\theta_{A_0} - \theta_{A(ss)}]\left[\frac{T_1}{T_1 - T_2} e^{-t/T_1} - \frac{T_2}{T_1 - T_2} e^{-t/T_2}\right] + \theta_{A(ss)}\right\}$$

$$+ \left\{\frac{(\theta_{A_0} - \theta_{A(ss)})\tau_2}{T_1 - T_2} (e^{-t/T_2} - e^{-t/T_1})\right\}. \quad (7.54)$$

Here we have an illustration of the superposition principle applicable to linear systems. The first bracket term is the response to the step function alone. The second, which can be obtained directly by taking $\theta_A = 0$ and $\dot{\theta}_A = (\tau_2/T_1 T_2)[\theta_{A(ss)} - \theta_{A_0}]$ at $t = 0$ in the complementary function (7.41), is the response to the impulse alone. When the two forcings are applied simultaneously, as in (7.27), the response is the sum of the responses to each forcing alone. Finally we note that (7.54) could have been obtained in still another way, i.e., by taking the derivative of (7.47) to get $\dot{\theta}_T$, multiplying it by τ_2, and adding the result to (7.47) to obtain θ_A in accordance with Eq. (7.24).

Let us now examine the Laplace-transform approach to the solution of (7.26) and (7.27). We first rewrite them in terms of our time constants to obtain:

$$T_1 T_2 \ddot{\theta}_T + (T_1 + T_2)\dot{\theta}_T + \theta_T$$

$$= BA_s F_{CO_2}{}^I(t) + \frac{BA_s MR}{\dot{V}_A} + \frac{MR}{Q} + A_i \quad (7.55)$$

$$T_1 T_2 \ddot{\theta}_A + (T_1 + T_2)\dot{\theta}_A + \theta_A$$

$$= \tau_2 \dot{F}_{CO_2}{}^I(t) + F_{CO_2}{}^I(t) + \frac{MR}{\dot{V}_A}. \quad (7.56)$$

Now, if our forcing is to be an $F_{CO_2}{}^I$ step function of magnitude $(F_{CO_2}{}^I)_1$, we interpret the right side of (7.55) to mean, as before, that at $t = 0$:

$$\theta_T = \frac{BA_sMR}{\dot{V}_A} + \frac{MR}{Q} + A_i \equiv \theta_{T_0}$$

and as $t \to \infty$, $\theta_T = \theta_{T_0} + BA_s(F_{CO_2}{}^I)_1 \equiv \theta_{T(ss)}$. Similarly, for (7.56), we have at $t = 0$, $\theta_A = MR/\dot{V}_A \equiv \theta_{A_0}$, and as $t \to \infty$, $\theta_A = \theta_{A_0} + (F_{CO_2}{}^I)_1 \equiv \theta_{A(ss)}$. Again, there are two ways of handling the constant terms θ_{T_0} and θ_{A_0} in solving (7.55) and (7.56) by transform methods. The hard way is to assign them as initial conditions on θ_T and θ_A while simultaneously adding them to the magnitude of the step forcing. The easy way is to ignore them until the solutions are obtained and then simply add them on. Let us choose the easy way. The normal transforms of (7.55) and (7.56) are then:

$$[T_1T_2s^2 + (T_1 + T_2)s + 1]\theta_T(s) = BA_sF_{CO_2}{}^I(s) \qquad (7.57)$$

$$[T_1T_2s^2 + (T_1 + T_2)s + 1]\theta_A(s) = (\tau_2s + 1)F_{CO_2}{}^I(s). \qquad (7.58)$$

Taking $F_{CO_2}{}^I(s)$ as a step function of magnitude $(F_{CO_2}{}^I)_1$ and rearranging into transfer function form yields:

$$\theta_T(s) = \left[\frac{1}{T_1T_2s^2 + (T_1 + T_2)s + 1}\right]\left[\frac{BA_s(F_{CO_2}{}^I)_1}{s}\right] \qquad (7.59)$$

$$\theta_A(s) = \left[\frac{1}{T_1T_2s^2 + (T_1 + T_2)s + 1}\right]\left[\frac{(F_{CO_2}{}^I)_1}{s} + \tau_2(F_{CO_2}{}^I)_1\right]. \qquad (7.60)$$

The first bracket term on the right, identical in both equations, is the system transfer function. The second bracket term is the forcing-function transform. In (7.60), the term $\tau_2(F_{CO_2}{}^I)_1$ is the transform of the impulse component of the forcing. However, it is clear that exactly the same term would have arisen had we omitted the impulse component and instead assigned an initial condition $\tau_2(F_{CO_2}{}^I)_1/T_1T_2$ on $\dot{\theta}_A$. This equivalent effect of an impulse forcing in terms of an initial condition on the $(n - 1)$th derivative has been noted previously.

As a first step in obtaining the inverse transforms of (7.59) and (7.60), let us write them in the "standard fraction form" as follows:

$$\theta_T(s) = \frac{1/T_1 T_2 [BA_s (F_{CO_2}{}^I)_1]}{s(s + 1/T_1)(s + 1/T_2)} \tag{7.61}$$

$$\theta_A(s) = \frac{1/T_1 T_2 (F_{CO_2}{}^I)_1 + \tau_2 s (F_{CO_2}{}^I)_1}{s(s + 1/T_1)(s + 1/T_2)}. \tag{7.62}$$

Partial fraction expansion by the methods of Chapter 3 is then very easily performed to yield:

$$\theta_T(s) = \frac{BA_s (F_{CO_2}{}^I)_1}{s} + \frac{BA_s (F_{CO_2}{}^I)_1 T_1}{(T_2 - T_1)(s + 1/T_1)} \\ - \frac{BA_s (F_{CO_2}{}^I) T_2}{(T_2 - T_1)(s + 1/T_2)} \tag{7.63}$$

$$\theta_A(s) = \frac{(F_{CO_2}{}^I)_1}{s} + \frac{(F_{CO_2}{}^I)_1 (T_1 - \tau_2)}{(T_2 - T_1)(s + 1/T_1)} \\ - \frac{(F_{CO_2}{}^I)_1 (T_2 - \tau_2)}{(T_2 - T_1)(s + 1/T_2)} \tag{7.64}$$

and the inverse transforms are at once apparent:

$$\theta_T(t) = BA_s (F_{CO_2}{}^I)_1 \left[1 + \frac{T_1}{T_2 - T_1} e^{-t/T_1} - \frac{T_2}{T_2 - T_1} e^{-t/T_2} \right] \tag{7.65}$$

$$\theta_A(t) = (F_{CO_2}{}^I)_1 \left[1 + \frac{T_1 - \tau_2}{T_2 - T_1} e^{-t/T_1} - \frac{T_2 - \tau_2}{T_2 - T_1} e^{-t/T_2} \right]. \tag{7.66}$$

Finally, it is clear that if we simply add θ_{T_0} to (7.65) and θ_{A_0} to (7.66) these equations become identical with (7.46) and (7.52).

We can now draw a block diagram for our isolated controlled system under the assumption that the forcing term is $F_{CO_2}{}^I$. Returning to Eqs. (7.57) and (7.58), we can rewrite them in the following form:

$$\theta_T(s) = \left[\frac{BA_s}{(T_1 s + 1)(T_2 s + 1)} \right] F_{CO_2}{}^I(s) \tag{7.67}$$

$$\theta_A(s) = \left[\frac{\tau_2 s + 1}{(T_1 s + 1)(T_2 s + 1)} \right] F_{CO_2}{}^I(s) \tag{7.68}$$

and the corresponding block diagram appears in Fig. 68. Note
that only the normal response is included in the diagram. We
can include the constant terms which appear in Eqs. (7.55)
and (7.56) by complicating this diagram somewhat, as in Fig. 69
Since these terms can be accounted for simply by a scale-shifting

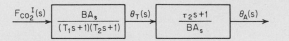

FIG. 68. BLOCK DIAGRAM OF ISOLATED CONTROLLED SYSTEM,
NORMAL RESPONSE

operation after the normal solutions are obtained, they are
usually omitted from explicit consideration either in the block
diagram or the transform equations. Thus we have already
noted (Chapter 2) that a constant disturbance effecting a servo
system or a set point other than zero for a regulator is generally
ignored in the theoretical treatment.

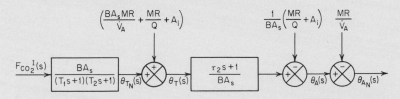

FIG. 69. ISOLATED CONTROLLED SYSTEM INCLUDING "BIAS"

Having solved (7.26) and (7.27) by both classical and Laplace
methods, let us now examine the behavior of $\theta_T(t)$ and $\theta_A(t)$ as
defined by Eqs. (7.46) and (7.52). To do this, it is most convenient to take $\theta_o = 0$ and $\theta_{ss} = 1$ in both equations to yield:

$$\theta_T = \frac{T_1}{T_2 - T_1} e^{-t/T_1} - \frac{T_2}{T_2 - T_1} e^{-t/T_2} + 1 \qquad (7.69)$$

and

$$\theta_A = \frac{T_1 - \tau_2}{T_2 - T_1} e^{-t/T_1} - \frac{T_2 - \tau_2}{T_2 - T_1} e^{-t/T_2} + 1. \qquad (7.70)$$

These equations are plotted in Fig. 70 for the normal set of
system properties defined in Table 5. Actually, the curves in

Fig. 70 were recorded directly from the output of an analog computer programmed to solve Eqs. (7.26) and (7.23), as shown in Fig. 71. The left-hand part of the circuit we recognize as identical with that of Fig. 33, and it solves the second-order equation (7.26) for θ_T. The rest of the circuit obtains θ_A from $\dot{\theta}_T$ and θ_T according to Eq. (7.23).

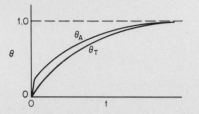

FIG. 70. RESPONSE OF θ_A AND θ_T TO A STEP FUNCTION OF $F_{CO_2}I$ IN ISOLATED CONTROLLED SYSTEM

Looking at Table 5, we note that the ratio T_2/T_1 is very large. This means that the first exponential term in (7.69) and (7.70) dies out comparatively rapidly, so that except for the first minute or so both equations behave as if they were first

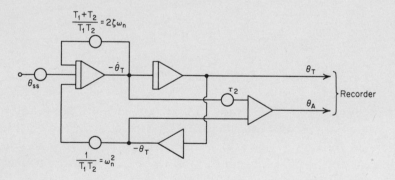

FIG. 71. ANALOG COMPUTER CIRCUIT FOR SOLUTION OF EQS. (7.26) AND (7.23)

order with time constant T_2. Even in the first minute, the first exponential term in (7.69) can be neglected since $T_1/(T_2 - T_1)$ is so small. This is not true of (7.70) however. These points are illustrated by Fig. 70 and by the corresponding semilogarithmic plots of $1 - \theta$ in Fig. 72. To obtain curves of this type experimentally, it would be necessary to maintain constant artificial ventilation and to measure $\theta_A(t)$ and $\theta_T(t)$ following a sudden change in inspired gas composition from air to, say, 5 percent CO_2.

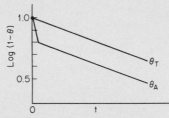

FIG. 72. SEMILOG PLOT OF RESPONSE OF θ_A AND θ_T TO A STEP FUNCTION OF $F_{CO_2}{}^I$

Frequency response. The graphic methods of Chapter 5 make it very easy to determine the steady-state response of θ_T and θ_A when $F_{CO_2}{}^I(t)$ is a sinusoid. In Fig. 73 we have drawn the asymptotic approximation of the Bode gain curve for each of the three component transfer functions, $[1/(T_1s + 1)]$, $[1/(T_2s + 1)]$, and

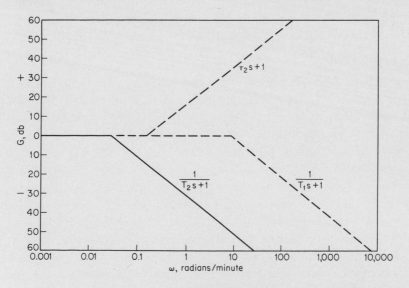

FIG. 73. BODE GAIN DIAGRAM FOR COMPONENTS OF ISOLATED CONTROLLED SYSTEM

$(\tau_2 s + 1)$ of Fig. 68. We have ignored the term BA_s, defined steady-state gains as unity for both θ_A and θ_T, and used normal values for the time constants. The corresponding phase curves appear in Fig. 74. Graphic addition of the appropriate transfer functions then yields the gain and phase curves for θ_T and θ_A

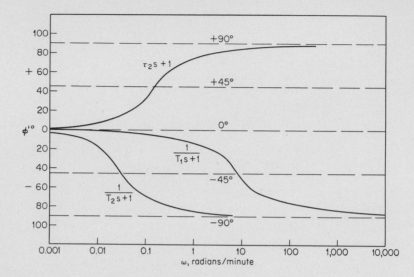

FIG. 74. BODE PHASE DIAGRAM FOR COMPONENTS OF ISOLATED CONTROLLED SYSTEM

drawn in Figs. 75 and 76. We can say that the θ_T response is flat out to the first corner frequency of 0.0304 radian/min. (or 0.0048 cpm), then attenuates 6 db/octave (like a first-order system) out to the second corner frequency of 8.34 radians/min. (or 1.33 cpm), and thereafter attenuates 12 db/octave (like a second-order system). The two corner frequencies are over two decades apart. The θ_A gain curve has three corners, two flat regions, and two regions where attenuation is 6 db/octave. If $F_{CO_2}{}^I$ were varied sinusoidally at respiratory frequencies (say 10–50/min., or about 60–300 radians/min.), θ_T would show practically no ac component (-90 db), whereas θ_A, although considerably attenuated (-40 db), would still show measurable oscillation.

Let us finally describe an electrical analog which has the same mathematical model as our isolated controlled system. The appropriate circuit is shown in Fig. 77. It consists of two dependently coupled "*RC* stages" provided with a voltage source $E(t)$ and a constant current source i. Except for the constants introduced by the CO_2 absorption curve, this electrical system has

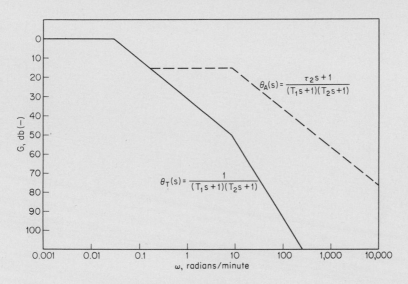

FIG. 75. BODE DIAGRAM SYNTHESIS OF θ_A AND θ_T GAIN CHARACTERISTICS

the same mathematical model as the biological system, with the following quantities being analogous:

$$R_1 = \frac{1}{\dot{V}_A} \qquad C_1 = K_A \qquad E(t) = F_{CO_2}{}^I(t) \qquad e_1(t) = \theta_A(t)$$

$$R_2 = \frac{1}{Q} \qquad C_2 = K_T \qquad i = MR \qquad e_2(t) = \theta_T(t).$$

Parametric forcing. As long as we force our isolated controlled system through $F_{CO_2}{}^I$, it behaves as a simple linear system with constant coefficients, and all of the neat mathematical machinery of Chapters 3, 4, and 5 can be directly applied as we have just seen. Appearing only in the unhomogeneous term on the right side, $F_{CO_2}{}^I$ is said to act as a direct forcing. But when the controlled system is forced by the respiratory center, the forcing term is $\dot{V}_A$. Looking at Eqs. (7.26) and (7.27), we see that $\dot{V}_A$ appears not only in the direct forcing term on the right, but also in the coefficients on the left. A forcing term which operates through these coefficients (or parameters) is called a parametric forcing.

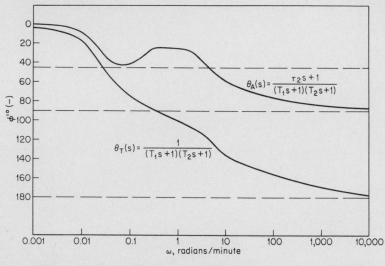

$$\theta_A(s) = \frac{\tau_2 s + 1}{(T_1 s + 1)(T_2 s + 1)}$$

$$\theta_T(s) = \frac{1}{(T_1 s + 1)(T_2 s + 1)}$$

FIG. 76. BODE DIAGRAM SYNTHESIS OF θ_A AND θ_T PHASE CHARACTERISTICS

Unfortunately, in parametric forcing, much of the mathematical simplicity disappears. Thus, even if our feedback loop were open and $\dot{V}_A$ were taken as an arbitrary function of time, the system equation, while still linear, would no longer have constant coefficients. Then, when we close the feedback loop, $\dot{V}_A$ becomes a function of θ_T, and the closed-loop equations are nonlinear. Nonlinearity arising from feedback through parameters is frequently encountered in biological systems. From the mathematical standpoint, it would be much nicer if our system responded to an $F_{CO_2}{}^I$ disturbance by reducing MR rather than by increasing $\dot{V}_A$, for then we would have a well-behaved linear regulator to which we could apply all of the analytical techniques outlined in previous chapters. But, admittedly, although this would simplify the mathematics, it would probably kill the mathematician!

Before turning to the closed-loop system, let us briefly examine the response of the isolated controlled system to parametric forcing by $\dot{V}_A$. We shall thus make a gradual transition from simple linear systems with constant coefficients through the still linear, but much less simple, systems with time-variable coeffi-

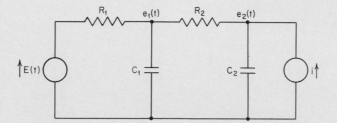

FIG. 77. ELECTRICAL ANALOG OF ISOLATED CONTROLLED SYSTEM

cients, to finally enter the complex world of nonlinear systems.

The simplest parametric forcing is that produced by a step change. Thus, if $\dot{V}_A$ steps from $\dot{V}_{A_0}$ to $\dot{V}_{A_1}$ at $t = 0$ and remains there, the parameters will be constant for $t > 0$ so that the problem becomes essentially identical with that of direct forcing. The only difference lies in establishing the initial conditions. We cannot do this by looking only at the second-order equations (7.26) and (7.27), but must also consider the more detailed information contained in the component equations (7.22) and (7.23). We note that up until $t = 0$ with $F_{CO_2}{}^I = 0$ and $\dot{V}_A = \dot{V}_{A_0}$, $\theta_T = (BA_sMR/\dot{V}_{A_0}) + (MR/Q) + A_i$, $\theta_A = MR/\dot{V}_{A_0}$, and $\dot{\theta}_A = \dot{\theta}_T = 0$. At $t = 0$, $\dot{V}_{A_0}$ steps instantaneously to $\dot{V}_{A_1}$, and, putting this value along with the above values for θ_A and θ_T into (7.22), we find that the effect of the parametric forcing, in addition to changing the values of the (constant) coefficients, is to impose an initial condition of $(MR/K_A)[1 - [\dot{V}_{A_1}/\dot{V}_{A_0}]$ on $\dot{\theta}_A$. This is reminiscent of the effect of the $F_{CO_2}{}^I$ impulse term in (7.27). Hence, for given values of the coefficients for $t > 0$, the dimensionless transient responses of θ_T and θ_A to step parametric forcing by $\dot{V}_A$ should be identical with those produced by direct forcing through $F_{CO_2}{}^I$. However, when we say "for given values of the coefficients for $t > 0$," we must remember that, unlike direct forcing by $F_{CO_2}{}^I$, parametric forcing by $\dot{V}_A$ changes the values of these coefficients at $t = 0$.

If $\dot{V}_A(t)$ is some other type of function, this simple treatment of parametric forcing no longer applies. Moreover, although Eqs. (7.26) and (7.23) are valid whether $\dot{V}_A$ is a constant or an arbitrary function of time, Eq. (7.27) must be replaced by a more complex one. Although the equations are still linear and the general solutions consist of the sum of a complementary function and particular integral as before, the details of the analytical techniques required to obtain them are too involved for us to consider. Instead, we shall use the analog computer to explore the frequency response of the system to parametric forcing. The analog circuit used will be described later. Over the frequency

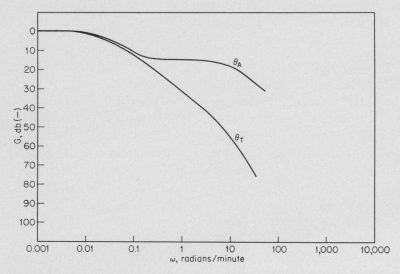

FIG. 78. BODE DIAGRAM FOR PARAMETRIC FORCING OF ISOLATED
CONTROLLED SYSTEM BY $\dot{V}_A$

range studied, the system output is essentially sinusoidal, and the Bode gain curve shown in Fig. 78 very closely resembles that for direct forcing in Fig. 74. In general, however, the output of a parametrically forced system will show parametric distortion, i.e., it will contain frequency components not present in the forcing. We will not consider this further but will turn instead to the controlling system and the closed-loop chemostat.

The controlling system and the closed-loop chemostat. According to our seventh assumption on page 114, the controlling system is to be a "simple proportional controller containing no dynamic elements." Equation (7.6) defines an empirical relationship between arterial pCO_2 and $\dot{V}_A$ known to hold in the steady state of CO_2 inhalation at rest. We now assume that a similar relationship can be written in terms of θ_T, and that it will hold for transient as well as steady-state operation. Accordingly, we write for the controlling-system equation:

$$\dot{V}_A = k_p[\theta_T - \theta_{T(i)}] + \dot{V}_{A(r)}, \tag{7.71}$$

where $\theta_{T(i)}$ is the set point and $\dot{V}_{A(r)}$ is the reference value. The gain of the proportional controller is k_p.

We can now draw a block diagram for the closed-loop dynamic chemostat (Fig. 79). Note that the disturbing quantity

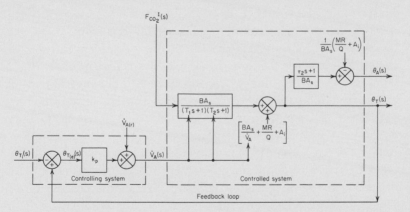

FIG. 79. BLOCK DIAGRAM OF DYNAMIC CHEMOSTAT (CLOSED-LOOP)

$F_{CO_2}{}^I(s)$ acts as a direct forcing to the controlled system, but that the controlling quantity $\dot{V}_A(s)$ forces the system parametrically. The latter is the source of the system nonlinearity as we can see by substituting (7.71) into (7.26) to obtain the following closed-loop equation for θ_T:

$$\alpha\ddot{\theta}_T + \beta\dot{\theta}_T + \gamma\dot{\theta}_T\theta_T + \theta_T{}^2 + \eta\theta_T = \lambda, \tag{7.72}$$

where

$$\alpha \equiv K_A K_T / k_p Q$$

$$\beta \equiv \frac{K_A + K_T BA_s}{k_p} - \frac{K_T \theta_{T(i)}}{Q} + \frac{K_T \dot{V}_{A(r)}}{k_p Q}$$

$$\gamma \equiv \frac{K_T}{Q}$$

$$\eta \equiv \frac{\dot{V}_{A(r)}}{k_p} - \theta_{T(i)} - BA_s F_{CO_2} I - \frac{MR}{Q} - A_i$$

$$\lambda = \frac{BA_s MR}{k_p} + \left[\frac{\dot{V}_{A(r)}}{k_p} - \theta_{T(i)} \right] \left[BA_s F_{CO_2} I + \frac{MR}{Q} + A_i \right].$$

Equation (7.72) is a nonlinear differential equation because it contains the second-degree terms $\dot{\theta}_T \theta_T$ and θ_T^2. Since $\dot{V}_A$ is a linear algebraic function of θ_T [Eq. (7.71)], the closed-loop equation for $\dot{V}_A$ will be identical in form with (7.71). A complex, nonlinear closed-loop equation can be derived for θ_A, but we shall not present it here.

No general analytical theory exists for the solution of nonlinear differential equations. Particular solutions can be obtained by numerical methods which, thanks to modern digital computers, are no longer prohibitively laborious and time-consuming. The analog computer offers another very convenient method of solution, and this is the tool we shall use here. There are many possibilities in designing an analog circuit to solve a particular problem. In many cases, the choice will be dictated by a desire for maximum convenience and flexibility. On the other hand, if only limited facilities are available, the circuit which requires the minimum number of components will be chosen. We could design a circuit for θ_T based directly upon Eq. (7.72) and introduce additional components to obtain $\dot{V}_A$ from (7.71) and θ_A from (7.23). Actually, if we had planned in terms of an analog computer study from the beginning and did not particularly care about the analytical form of the combined equations, we would introduce the computer very early. Thus we would have designed our circuit to solve (7.22) and (7.23) simultaneously with $\dot{V}_A$ a constant for the open loop or with $\dot{V}_A$ defined by (7.71) for the closed loop. We could thus bypass all of the manipulations required to obtain the combined equations, (7.26) and (7.27), as well as their solutions by classical or Laplace techniques, and we would not have had to derive the composite closed-loop equation (7.72) at all!

Such an analog circuit is shown in Fig. 80 where we have neglected scaling for simplicity. The only unfamiliar component

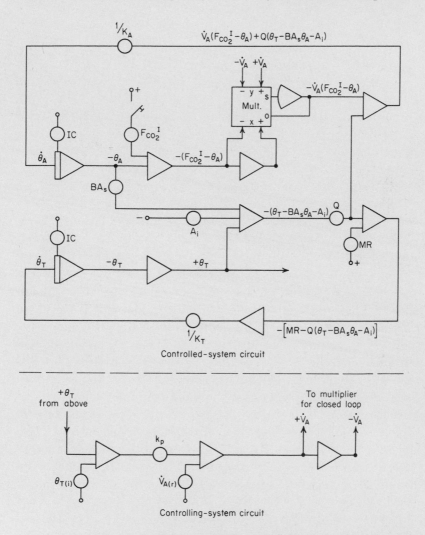

Controlled-system circuit

Controlling-system circuit

FIG. 80. ANALOG COMPUTER CIRCUIT FOR SOLUTION OF DYNAMIC CHEMOSTAT EQUATIONS

is the "quarter square multiplier" at the top which performs the operation $\dot{V}_A(F_{CO_2}{}^I - \theta_A)$ called for by Eq. (7.22) under the assumption that, in general, $\dot{V}_A$ will be a function of time. This circuit will generate solutions for all of the open-loop problems already considered and in addition will solve the

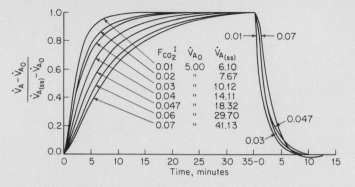

FIG. 81. THEORETICAL VENTILATION TRANSIENTS FOR VARIOUS VALUES OF $F_{CO_2}{}^I$

From F. S. Grodins, J. S. Gray, K. R. Schroeder, A. L. Norins, and R. W. Jones, "Respiratory responses to CO_2 inhalation: a theoretical study of a nonlinear biological regulator," *J. Appl. Physiol.* 7 (1954): 283.

closed-loop problems yet to be considered. All we need do is take the $\pm\dot{V}_A$ input to the multiplier from the appropriate source. Thus, for $F_{CO_2}{}^I$ forcing in the open loop, $\dot{V}_A$ is simply a constant. For $\dot{V}_A$ step forcing in the open loop, we change $\dot{V}_A$ from one constant value to another by throwing a switch. For sinusoidal forcing of $\dot{V}_A$ in the open loop, $\dot{V}_A$ is taken from a sine-wave generator (which may be the solution of the harmonic oscillator programmed within the same computer). Finally, for the closed loop, $\dot{V}_A$ is the output of the controlling-system circuit at the bottom of the figure which takes its input from θ_T. The dynamic behavior of θ_T, θ_A, and $\dot{V}_A$ (along with others if desired) can all be recorded simultaneously. The great convenience of the analog computer for studying problems of this sort is therefore apparent.

Let us now examine the closed-loop response to a step function $F_{CO_2}{}^I$ disturbance and see whether or not it resembles the observed behavior (Fig. 66) which prompted the development of this dynamic model in the first place. We shall plot the behavior of ventilation and arterial pCO_2* for both the "on transient" [i.e., after stepping $F_{CO_2}^I$ from zero to $(F_{CO_2}^I)_1$] and the "off transient" [i.e., after stepping $F_{CO_2}^I$ from $(F_{CO_2}^I)_1$ to zero]. For a linear system, these two transients would be iden-

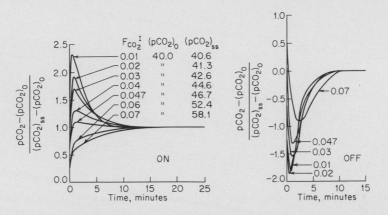

FIG. 82. THEORETICAL pCO_{2A} TRANSIENTS FOR VARIOUS VALUES OF $F_{CO_2}^I$

From F. S. Grodins, J. S. Gray, K. R. Schroeder, A. L. Norins, and R. W. Jones, "Respiratory responses to CO_2 inhalation: a theoretical study of a nonlinear biological regulator," *J. Appl. Physiol.* 7 (1954): 283.

tical, but this is not the case in our system, as we shall see. For convenience, we shall plot all responses in dimensionless form. Using normal values for the system properties, the theoretical response of $\dot{V}_A$ for various values of $(F_{CO_2}^I)_1$ are shown in Fig. 81, and the corresponding behavior of pCO_{2A} is shown in Fig. 82. It is apparent that most of the unique features of the prototype behavior illustrated in Fig. 66 are reproduced by the model.

Thus, in both during the "on transient," ventilation rises relatively slowly to reach its new steady state without overshoot, whereas pCO_{2A} rises much more rapidly, overshoots slightly, and then gradually falls to its new steady-state value. Again, in both during the "off transient," ventilation falls more rapidly than it rose and does not undershoot. Arterial pCO_2, on the contrary, falls very rapidly and undershoots markedly before

* Arterial $pCO_2 = B\theta_A$.

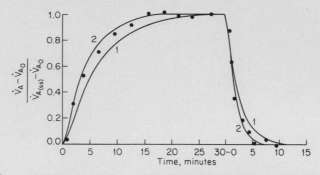

FIG. 83. EXPERIMENTAL AND THEORETICAL VENTILATION
TRANSIENTS FOR 5 PERCENT CO_2

• = Grodins' data, for six subjects. Smooth curves are theoretical responses for
normal system properties (1) and for $K_T = 30l$ (2). From F. S. Grodins, J. S. Gray,
K. R. Schroeder, A. L. Norins, and R. W. Jones, "Respiratory responses to CO_2
inhalation: a theoretical study of a nonlinear biological regulator," *J. Appl. Physiol.*
7 (1954): 283.

rising gradually to its control level. The agreement is even more
encouraging when experimental data for various values of $(F_{CO_2}I)_1$
are superimposed upon the theoretical curves in Figs. 83 and 84.
We can say that the mathematical model appears to be a useful
description of the biological prototype at least for this particular
type of disturbance.

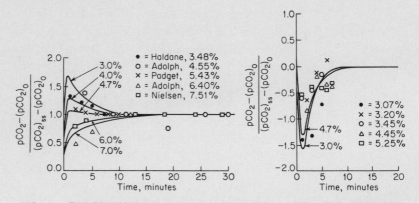

FIG. 84. EXPERIMENTAL AND THEORETICAL pCO_{2A} TRANSIENTS
FOR VARIOUS VALUES OF $F_{CO_2}I$

From F. S. Grodins, J. S. Gray, K. R. Schroeder, A. L. Norins, and R. W. Jones,
"Respiratory responses to CO_2 inhalation: a theoretical study of a nonlinear biological regulator," *J. Appl. Physiol.* 7 (1954): 283.

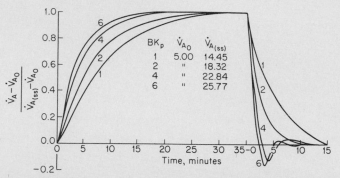

FIG. 85. THEORETICAL VENTILATION RESPONSES TO 5 PERCENT
CO_2 FOR VARIOUS VALUES OF CONTROLLER GAIN k_p

From F. S. Grodins, J. S. Gray, K. R. Schroeder, A. L. Norins, and R. W. Jones,
"Respiratory responses to CO_2 inhalation: a theoretical study of a nonlinear biological regulator," *J. Appl. Physiol.* 7 (1954): 283.

The analog computer is also extremely valuable in exploring
the effects of "abnormal" system properties upon the responses.
In this way we can rapidly identify those properties which are
most critical in determining system behavior. We shall not go
into this aspect in any detail but simply give one illustrative
example. Thus the curves in Figs. 85 and 86 show the effect of
alterations in controller gain k_p on the transient response to
5 percent CO_2.

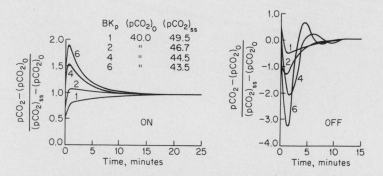

FIG. 86. THEORETICAL pCO_{2A} RESPONSES TO 5 PERCENT CO_2 FOR
VARIOUS VALUES OF CONTROLLER GAIN

From F. S. Grodins, J. S. Gray, K. R. Schroeder, A. L. Norins, and R. W. Jones,
"Respiratory responses to CO_2 inhalation: a theoretical study of a nonlinear biological regulator," *J. Appl. Physiol.* 7 (1954): 283.

How would the chemostat behave if it responded to an error in θ_T by manipulating MR instead of $\dot{V}_A$? Let us suppose that the controlling-system equation were the following:

$$MR = k_p[\theta_{T(i)} - \theta_T] + MR_{(r)}. \tag{7.73}$$

From Chapter 4, we know that closing this proportional feedback loop around the linear second-order system represented by Eq. (7.26) will not change the form of the equation but will reduce the damping ratio and increase the natural frequency. Using the analog circuit of Fig. 71 it is very easy to explore the responses of this closed-loop system to a step function $F_{CO_2}{}^I$ dis-

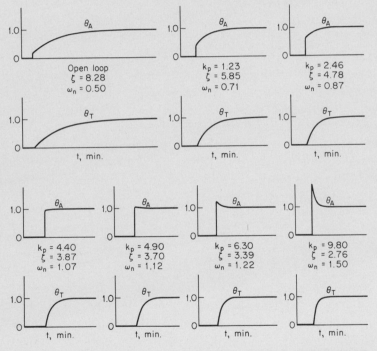

FIG. 87. BEHAVIOR OF HYPOTHETICAL LINEAR CHEMOSTAT

turbance for various values of controller gain k_p. The results are summarized in Fig. 87. Note that, qualitatively, the effects of increasing k_p in the linear system are quite similar to those of increasing k_p in the nonlinear system (Fig. 85). In the open-loop system ($k_p = 0$), both θ_A and θ_T are overdamped. As

k_p increases, both responses become faster and θ_A begins to overshoot while θ_T is still overdamped. The off transients are of course identical for this linear system and are not shown.

What then are the essential differences between the responses of the linear and the nonlinear systems? There are two obvious ones. First, the on and off transients are identical in the former but not in the latter. Second, the form of response is independent of the magnitude of the $F_{CO_2}^I$ step function in the linear system but not in the nonlinear one. This dependence of the form of response upon the amplitude and/or range of the forcing function is characteristic of nonlinear systems. Nevertheless, as we have seen, certain features of behavior are quite similar for both systems.

Recent developments. In recent years, this dynamic chemostat has been refined and extended. Thus the lumped tissue reservoir has been separated into two parallel components (brain and not brain) and provision has been made to vary brain blood flow as a function of arterial pCO_2.* The respiratory cycle has also been included and oscillatory arterial pCO_2 signals arising from it have been used to trigger an integral controller in an attempt to explain the hyperpnea of exercise.† Finally, the model has been extended to include anoxemia and metabolic disturbances in acid-base balance. These developments are beyond the scope of this monograph.

Summary

During World War II, a steady-state feedback model was developed to explain the respiratory responses to CO_2 inhalation, arterial anoxemia, and metabolic disturbances in acid-base balance. The controlled-system outputs were taken as arterial pCO_2, (H^+), and pO_2, and these were fed back as input to the controlling system. The output of the latter was pulmonary ventilation. The model has been a very fruitful one.

* J. G. Defares, H. E. Derksen, and J. W. Duyff, "Cerebral blood flow in the regulation of respiration," *Acta Physiol. et Pharmacol. Neerl.* 9 (1960): 327.

† F. S. Grodins and G. James, "Mathematical models of respiratory regulation," *Ann. N. Y. Acad. Sci.*, in press.

During the onset of and recovery from inhalation of CO_2, ventilation and arterial pCO_2 dissociate in a way which suggests that the latter is not the actual input to the controller. A dynamic model was developed to explain this behavior on the assumption that the controller input was tissue (respiratory center) CO_2 concentration. The differential equation of the controlled system was of the familiar linear, second-order variety encountered in previous chapters. However, when the loop was closed something new was encountered, i.e., feedback through parameters yielding a nonlinear closed-loop system. Some of the features of this system were examined with the help of an analog computer.

The difficulties encountered in this chapter are typical of biological systems. We shall meet them again when we consider the cardiovascular system in Chapter 8.

CHAPTER 8

The Cardiovascular Regulator

THE mammalian cardiovascular system is a very complex hydrodynamic system whose many parameters are continuously operated upon by neural and humoral controlling signals. How shall we approach this formidable puzzle from the standpoint of control-system theory? Let us begin by considering an identification problem,* i.e., what are the controlled quantities which this regulator attempts to stabilize and what variables does it manipulate in order to do so? This problem can be approached in various ways, but since we have just considered the respiratory chemostat, it is perhaps natural to first look at the cardiovascular system in a somewhat similar way. Let us therefore seek similarities between the two systems and see how far these analogies will take us.

A steady-state cardiovascular chemostat

If a person familar with the steady-state respiratory chemostat of Chapter 7 took a first look at the cardiovascular system, he could at once identify certain analogous features of a hypothetical cardiovascular chemostat. Thus, if he said that the respiratory system ventilates the lung reservoir with fresh air at a rate required to maintain alveolar (arterial) pO_2 and pCO_2 at or near normal levels, he might also say that the cardiovascular system perfuses the lumped-tissue reservoir with fresh blood at a rate required to keep tissue (mixed venous) pO_2 and pCO_2 at or

* This term has been used in a somewhat different sense elsewhere, e.g., E. Mishkin and L. Braun, Jr., *Adaptive Control Systems* (New York, McGraw-Hill Book Company, Inc., 1961).

near normal levels. From this standpoint, the controlled variables would be tissue (mixed venous) pO_2 and pCO_2, and the manipulated variable would be cardiac output. The block diagrams of the two chemostats would then be very similar. In fact, if we ignore the details of the CO_2 and O_2 absorption curves and substitute volumetric contents for tensions in both air and blood, the essential similarity of the two controlled-system blocks becomes very obvious (Fig. 88). We have omitted the H^+ loop for simplicity.

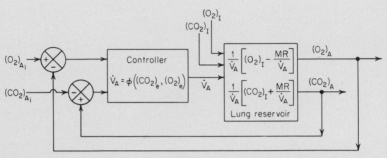

(a)

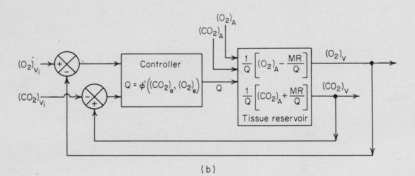

(b)

*FIG. 88. (a) BLOCK DIAGRAM FOR THE RESPIRATORY CHEMOSTAT;
(b) AN ANALOGOUS CIRCULATORY CHEMOSTAT*

But now let us note a difference in emphasis in the approach to these two apparently very similar regulators. Respiratory problems, clinical as well as physiological, have long revolved about disturbances in alveolar (arterial) gas tensions and blood acid-base balance. Thus, asphyxia, mountain sickness, hypox-

emia and/or hypercapnia associated with lung disease, hyperventilation tetany, and the air hunger of diabetic or nephritic
acidosis all emphasized the relationship between ventilation
and chemical composition of the blood. Moreover, ventilation
was very easy to measure, and receptors capable of producing
changes in ventilation were known to be located directly in the
path of the system output, arterial blood. Hence a model such
as Gray's steady-state chemostat, whose controller equation
side-stepped all of the neuro-muscular-mechanical details involved in transforming arterial composition into alveolar ventilation, was and is extremely valuable in understanding many of
the respiratory control problems considered of primary importance by physiologists and clinicians alike. But what about
the circulatory chemostat?

Although stagnant anoxia was early recognized as one result
of reduced tissue blood flow, the possible dependence of cardiac
output upon mixed venous chemistry has never received much
emphasis in discussions of cardiovascular regulation.* There are
many reasons for this. One stems simply from technical difficulties. Whereas both the input $(\dot{V}_A)$ and the output (arterial
composition) of the controlled system of the respiratory chemostat in Fig. 88a are easy to measure in intact man, neither of the
corresponding quantities (Q and mixed venous composition) of
the circulatory chemostat in Fig. 88b is. As a result physiologists and clinicians alike have long emphasized the importance
of cardiovascular variables which do not even appear in our
diagram, e.g., arterial and venous pressures and heart rate. Of
course, we do not really know whether our diagram is correct,
for we have simply drawn it as an analog of the respiratory
chemostat. The part in question is not the isolated tissue reservoir, for the equations describing it, like those of the lung
reservoir, depend only upon basic physical conservation principles. What we do not know is whether Q actually is manipulated in response to errors in mixed venous composition as our

---

* A notable exception is the concept expressed by W. F. Hamilton which is
qualitatively very similar to the one to be developed here: "Physiology of the
cardiac output," *Circulation* 8 (1953): 527. "Role of Starling concept in regulation
of normal circulation," *Physiol. Revs.* 35 (1955): 161.

diagram implies. If it is, we might expect to find a chemoceptor somewhere in the mixed venous blood stream, i.e., somewhere between the right atrium and the pulmonary capillaries. Unfortunately, despite much recent search (which, incidentally, was stimulated by another persistent problem—exercise hyperpnea) no one has yet been able to conclusively demonstrate such a receptor. Before discarding our diagram too hastily, however, let us ask whether information about mixed venous composition might be obtained indirectly by monitoring some other quantity. An obvious candidate is systemic arterial pressure, for we know there are baroceptors which respond to it.

If we neglect venous pressure and regard the systemic arterial-arteriolar system as a lumped, linear RC network, then it is clear that the systemic arterial pressure P_{AS} is simply the product of cardiac output Q and systemic peripheral resistance R_S:

$$P_{AS} = QR_S. \tag{8.1}$$

Let us now isolate the controlled system of Fig. 88b and add a "mechanical section" to it as in Fig. 89. Now, in particular

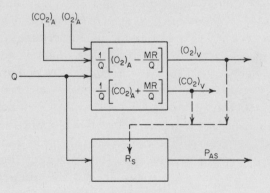

FIG. 89. BLOCK DIAGRAM TO SHOW DEPENDENCE OF P_{AS} ON VENOUS COMPOSITION

systemic vascular beds, it is known that the arteriolar resistance varies with local tissue (venous) composition as a direct chemical effect independent of nerves. We have indicated this action for the lumped system by the dashed arrows in Fig. 89. It is now apparent that if we monitor P_{AS} in the system of Fig. 89, we are also obtaining information about mixed venous composition. Hence we need not yet abandon the concept of a circulatory chemostat. We might have to complicate its block diagram (Fig. 88), however, by indicating that the feedback from mixed venous composition is indirect via P_{AS} as shown in Fig. 89.

Let us now point out another important difference between the two chemostats. We usually think of the lung as a single organ and do not hesitate to lump its many parallel components into a single one whose input is $\dot{V}_A$ and whose output is mixed alveolar (arterial) composition. After all, the lung comprises a passive element all portions of which have the same function. Although some nonuniformities do occur and may become exaggerated and important under certain conditions, purposeful redistribution of $\dot{V}_A$ does not appear to be a major feature of normal operation. But how different is the circulatory chemostat! The various parallel components of the tissue reservoir are active elements with very different functions, and so we suspect that each might have a chemostat of its own. Thus the differences between parallel circuits which are of secondary importance for the normal respiratory chemostat become of major interest in the cardiovascular system.

Let us therefore divide our tissue reservoir into a number of parallel circuits. In so doing we shall recognize that the distribution of the total cardiac output among the several circuits depends upon the individual circuit resistances. A block diagram for this new isolated controlled system appears in Fig. 90. For simplicity, we have retained only a single chemical agent, O_2, without necessarily implying that it is the only or even the most important one. Again it is apparent that P_{AS} is a measure of $(O_2)_V$. It is also apparent that if P_{AS} were kept constant (or nearly so) by manipulating Q, then the flow through each parallel circuit would depend only upon the resistance of that circuit and thus upon its chemical output. In effect, each parallel element as well as the entire lumped-tissue block would behave like the respiratory chemostat! This leads us directly to our next problem.

We have seen that P_{AS} is an indirect measure of $(O_2)_V$ and have noted that if Q were appropriately manipulated in response to errors in P_{AS}, the entire lumped system as well as each of its parallel components would function as chemostats. A crucial question, then, is whether or not Q is actually manipulated in response to errors in P_{AS}. Strangely enough (or perhaps not so strangely because technical difficulties are formidable) existing

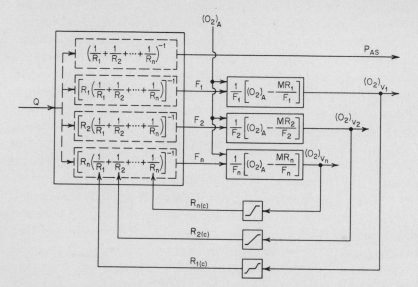

FIG. 90. *ISOLATED CONTROLLED SYSTEM OF STEADY-STATE CARDIOVASCULAR CHEMOSTAT*

experimental evidence on this point is quite meager and fails to give an unequivocal answer. Although there appears to be little question that heart rate is manipulated in response to errors in P_{AS}, this only suggests, but does not prove, that corresponding changes in Q occur. Much more experimental study is needed here.

If errors in P_{AS} do not result in changes in Q (exclusively), what else might be manipulated to regulate P_{AS}? There seems to be no question that R_S is manipulated through vasomotor nerves, and this feature of the system has long been emphasized. Moreover, there appears to be considerable difference in the susceptibility of various systemic circuits to such manipulation. Those tissues whose arterioles are most easily dilated by direct feedback of their chemical output are least susceptible to manipulation by vasomotor nerves (brain, heart, skeletal muscle) and vice versa (skin, gastrointestinal tract, kidney). What does this mean with reference to our concept of a cardiovascular chemostat? It certainly does not destroy it, but it does imply that the system has another degree of freedom in its operation. Instead of satisfying all tissues simultaneously by appropriate manipulation of Q, it can alternatively choose to preferentially supply the demands of some tissues at the expense of others. Thus the control system is clearly an adaptive one, i.e., it can adjust its controlling operations to changes in controlled-system conditions. It would appear that the brain and the heart (as well as skeletal muscle during exercise) are favored, an arrangement having obvious survival value.

Let us now summarize our concept of the steady-state cardiovascular chemostat, as we have so far developed it, in the block diagram of Fig. 91. The transfer functions of certain blocks have been represented by graphs, and these should be regarded only as rough qualitative guesses. What we wished them to show was that the arterioles of the various systemic vascular beds respond differently to local chemical feedback and to vasomotor nerves. How might such a system work? Suppose that $(O_2)_A$ were reduced by breathing a low-oxygen mixture. Our diagram says that this should first reduce all of the n venous (O_2) values by the same amount. This, via the direct chemical feedback loops, would reduce $R_1, R_2 \ldots R_n$ by various amounts.[*] The resulting fall in R_S would reduce P_{AS} and this, in turn, would increase Q and/or increase arteriolar resistance in selected areas.

Does this really happen when one breathes a low-oxygen mixture? There appears to be general agreement that cardiac output does increase in arterial anoxemia, perhaps by 50–75 percent at an altitude of 18,000 feet, and that mixed venous (O_2) is low. The changes reported in P_{AS} are somewhat less clearcut, some groups reporting a fall, others essentially no change, and still others an increase. The latter seems to occur only in

---

[*] We note the possibility that this dialator action of low (O_2) may be opposed by the constrictor action of hypocapnia and alkalemia, but will not pursue this any further.

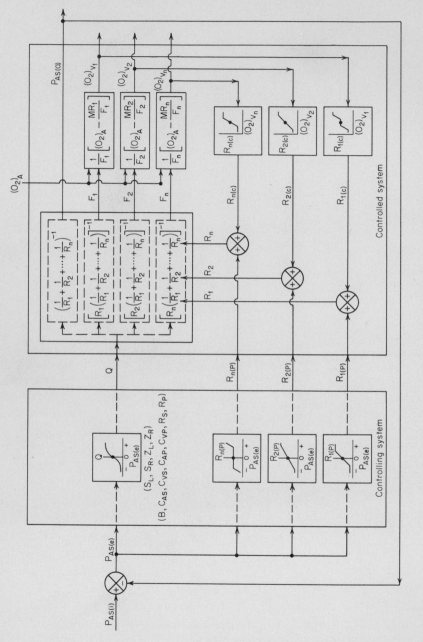

FIG. 91. BLOCK DIAGRAM OF THE STEADY-STATE CARDIOVASCULAR CHEMOSTAT

rather severe grades of anoxemia. Little is known about the distribution of the cardiac output in anoxemia. Note that our scheme assigns no important role to the carotid and aortic chemoceptors in the response of cardiac output to arterial anoxemia. This is consistent with recent work showing that these receptors are not responsible for the tachycardia which occurs. Does this scheme also explain cardiovascular responses to CO_2 inhalation and metabolic disturbances in acid-base balance? We really don't know whether it does or not because the details of the cardiovascular responses to these forcings have not been adequately studied. Inhalation of 5–7 percent CO_2 markedly reduces cerebral vascular resistance and increases cerebral blood flow, but the response of cardiac output has not been clearly established. In extreme degrees of anoxemia and hypercapnia, central chemoceptors not shown in our diagram become important and induce very marked arteriolar constriction in the skin, gastrointestinal tract, and kidneys. Our model could explain the increase in Q which occurs in anemia on a basis similar to that for arterial anoxemia, but here the accompanying reduction of blood viscosity may also play a role. Finally, we may ask whether our model can account for the cardiac output response to exercise.

Just as our respiratory chemostat model in Chapter 7 failed to account for exercise hyperpnea, so does our cardiovascular chemostat fail to explain the response of cardiac output. In the former case, the difficulty was that there was no (or minor) steady-state error in pCO_2, pO_2, or (H^+) during moderate exercise. In the latter case, the difficulty is that although there appears to be a steady-state error in P_{AS}, it is in the wrong direction, i.e., P_{AS} is usually said to be elevated in exercise! Here is a situation where P_{AS} does not seem to give us a faithful index of $(O_2)_V$, for the two move in opposite directions. The same sort of ad hoc hypothesis can be invoked to extend the application of both respiratory and circulatory chemostats to include exercise. What we can assume is that there exists a stimulus which is directly proportional to metabolic rate and that this stimulus simply adds to the respiratory controller equation (7.1) and "resets" $P_{AS(i)}$ in the cardiovascular regulator. No one has yet identified this stimulus (i), but over the years very similar guesses have been made for both chemostats, e.g., cortical radiation (under one name or another) and peripheral receptors (kinetic, thermo-, metabolo-, ergo-) in exercising muscles. In the case of the cardiovascular system, it has also been held that a mechanical factor, the so-called skeletal muscle pump, plays a major role. We do not wish to get involved in this problem any further at the moment, but will turn instead to another matter.

Let us suppose that we knew all of the transfer functions in Fig. 91, including the direct chemical effect of $(CO_2)_{V_{1-n}}$ and $(H^+)_{V_{1-n}}$ on R_{1-n} and that this chemostat satisfactorily accounted for the normal responses to arterial anoxemia, CO_2 inhalation, and metabolic disturbances in acid-base balance. Would we be as happy with the insight that this scheme provides for the cardiovascular system as we are with that provided for the respiratory system by Gray's model? The answer seems to be no and the reasons for this appear to be several.

Although we would like to know the details of the transformation from chemical stimulus to $\dot{V}_A$ for the respiratory pump, we do not regard their omission as a major defect. But this is hardly true for the cardiac pump! In fact, in terms of Fig. 91, we could say that a major concern of cardiovascular physiologists has been, and still is, the investigation of the exact way in which the function, $Q = \phi(P_{AS_{(o)}})$, depends upon the *mechanical* parameters of the heart and circulation. Why should this be so? It is probably because the major clinical problems concerning the cardiovascular system have always emphasized mechanical rather than chemical forcings, e.g., the effect of low blood volume (hemorrhage and shock), cardiac valvular lesions, myocardial failure, and arteriolosclerosis upon heart size, cardiac output, and vascular pressures. From the control standpoint, the adaptive behavior of the system is most clearly evident in response to certain mechanical forcings. Hence any control scheme which by-passes all mechanical details of the system cannot be regarded as very satisfactory. Let us therefore attempt to remedy this defect. In so doing, we shall become involved in system dynamics.

The mechanical cardiovascular system

The mechanical cardiovascular system has been a favorite subject for mathematical analysis ever since the work of Leonard Euler in 1775. Perhaps the names most familiar to contemporary physiologists in this area are those of Otto Frank, who with his many students dominated the field for the first 30 years of the current century,* and Womersley, McDonald, and Taylor who have made many significant contributions during the present decade.† Although these analyses have a high degree of mathematical sophistication, they treat only the most obvious feature of the circulation, i.e., the systemic arterial pulse. Dealing with an isolated arterial system forced by an arbitrary flow source (the left ventricle), mathematical models have been developed which vary in complexity from a lumped, linear first-order system (Frank's Windkessel) to the partial differential equations of hydrodynamics applied to blood flow in branched, distributed elastic tubes.

Although it is obvious that the distributed nature of the elastic vascular system with its attendant wave phenomena must be included in any complete theoretical description of the cardiovascular regulator, these analyses are not particularly useful for our present purpose. We are interested in a different sort of model, i.e., one which describes the interdependence of arterial pressures, venous pressures, and cardiac output in the closed-circuit cardiovascular system. Although frequently discussed in qualitative fashion, quantitative models of this sort have not attracted much attention (Van Harreveld, Guyton).‡ Let us now develop such a model in detail.

The heart-circuit model. In essence, this system consists of two pumps (right and left hearts) and two circuits (pulmonary and systemic) arranged in series. Each heart has two chambers,

---

* A. Aperia, "Hemodynamical studies," *Skand. Arch. Physiol.* 83 (1940), Suppl. 16.
† D. A. McDonald, *Blood Flow in Arteries* (London, E. Arnold, 1960).
‡ A. Van Harreveld and O. W. Shadle, "On hemodynamics," *Arch. intern. physiol.* 54 (1951): 165; A. C. Guyton, "Determination of cardiac output by equating venous return curves with cardiac response curves," *Physiol. Revs.* 35 (1955): 123.

atrium and ventricle, the exit and entrance of the latter being provided with flutter valves which ensure unidirectional flow through the heart. Each circuit consists of a bewildering array of branching elastic arteries and veins, muscular arterioles, and intricate capillary networks. For present purposes we shall lump each complex distributed circuit into three components: a single elastic artery, a single elastic vein, and a single arteriolar resistance. Each cardiac atrium will be lumped with its appropriate vein (Fig. 92).

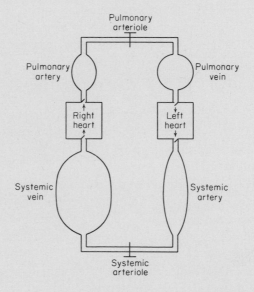

FIG. 92. THE HEART-CIRCUIT PROTOTYPE

Now the most important characteristic of this isolated mechanical system for our purposes is that the cardiac outputs (Q_R and Q_L) are dependent upon circuit pressures, and the circuit pressures in turn are dependent upon cardiac outputs. By virtue of this mechanical feedback, the system manifests self-regulation, i.e., if some disturbance should upset the equality of Q_R and Q_L, pressures in the circuits are automatically readjusted in just the way required to reestablish this equality. This is a very important stabilizing feature as anyone who has had experience with cardiac by-pass systems will testify. Now we should be able to identify a group of heart and circuit properties (independent variables) whose values determine the levels of Q_R, Q_L, and circuit pressures. From the more general point of view, it is these properties which are actually operated upon by the controlling system to establish the values of Q, F_{1-n}, and circuit pressures. We must now consider this problem in more specific terms.

The isolated heart.[*] The basic prototype of interest is the isolated ventricle preparation. Here the complex physiological circuit is replaced by a controllable artificial system. The investigator can arbitrarily set the venous filling pressure, the arterial load pressure, and the cardiac frequency, and thus can study the influence of each upon cardiac output and ventricular volumes. We shall analyze the steady-state operation of this isolated ventricle in terms of a filling process and an emptying process. Starting with the former, we suppose that we can isolate a single ventricular diastole. At its beginning, the ventricle has a certain arbitrary residual volume, v_r. It then fills for an arbitrary length of time t, at the end of which its volume will be the diastolic volume v_d.

We wish to define the dependence of v_d upon v_r, t, and the mechanical forces which drive and oppose filling. This is a formidable task and no really satisfactory solution is yet available. Let us use a very simple model as a first approximation: we will neglect any atrial contribution to ventricular filling and assume that (1) ventricular relaxation occurs instantaneously at the end of systole and is thus complete before filling begins, (2) filling of the relaxed ventricle is a first-order linear process driven by a constant venous pressure and opposed by viscoelastic forces generated within the heart, and (3) the unstressed volume of the relaxed ventricle is zero.

These assumptions turn out to yield a model which is very useful despite its neglect of certain obvious details. The corresponding differential equation is:

$$R\dot{v}_d + \frac{1}{C}v_d = P_V, \qquad (8.2)$$

---

[*] F. S. Grodins, "Integrative cardiovascular physiology: a mathematical synthesis of cardiac and blood vessel hemodynamics," *Quart. Rev. Biol.* 34 (1959): 93.

where R is the total viscous resistance to filling, C is the compliance of the relaxed ventricle, v_d is diastolic volume, and P_V is venous filling pressure. We recognize this equation as being identical in form with (2.5) derived for the spring-dashpot system of Chapter 2. Taking $\tau = RC$, the normal Laplace transform becomes:

$$(\tau s + 1)v_d(s) = CP_V(s), \qquad (8.3)$$

or in transfer-function form:

$$v_d(s) = \left(\frac{C}{\tau s + 1}\right)P_V(s), \qquad (8.4)$$

and the block diagram appears in Fig. 93. We know that the

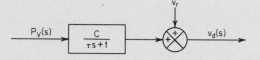

FIG. 93. BLOCK DIAGRAM FOR VENTRICULAR FILLING PROCESS

solution of (8.2) will be a single exponential of the type encountered in Chapter 4, and we can write it down at once:

$$v_d = (v_r - CP_V)e^{-t/\tau} + CP_V. \qquad (8.5)$$

Equation (8.5) describes the behavior of $v_d(t)$ during the filling process.

Turning now to the emptying process, we suppose that we can isolate a single ventricular systole. At its beginning, the ventricle has a certain diastolic volume v_d. It then ejects a certain stroke volume v_s against an arbitrary arterial load pressure P_A thus leaving behind when systole is over a certain residual volume v_r. We wish to define the dependence of v_s (and thus v_r) on v_d and P_A. Once more we face a formidable problem whose solution in terms of the fundamental properties of heart muscle and ventricular geometry is yet to be attained. Once more we shall use a very simple model, this time based on Starling's empirical "Law of the Heart." In so doing we neglect the kinetics of the process and look only at its beginning and end. Restricting our model to the region of compensation, we assume that ventricular useful stroke work, w_s, is directly proportional to diastolic volume:

$$w_s = Sv_d, \tag{8.6}$$

where the proportionality constant S will be called the strength of the ventricle. Neglecting any kinetic work, we can equate w_s to the product $P_A v_s$, so that (8.6) becomes:

$$P_A v_s = Sv_d \tag{8.7}$$

and
$$v_s = \frac{S}{P_A} v_d \qquad \text{for } \frac{S}{P_A} \leqq 1; \tag{8.8}$$

if > 1, then $v_s = v_d$,

$$v_r = v_d(1 - S/P_A) \qquad \text{for } \frac{S}{P_A} \leqq 1; \tag{8.9}$$

if > 1, then $v_r = 0$.

Equations (8.8) and (8.9) describe the end result of the emptying process but do not define its kinetics.

If we solve (8.5), (8.8), and (8.9) simultaneously for v_s, we can define the steady-state operation of the isolated ventricle. To do this, we will assume that systole has a constant duration of 0.2 sec and that the filling time t in (8.5) is $(60/f - 0.2)$ sec, where f is cardiac frequency in cycles/minute. To simplify our notation, we define:

$$K \equiv \exp\left[-(60/f - 0.2)/RC\right] \tag{8.10}$$
and
$$A = 1 - K. \tag{8.11}$$

We can then obtain the following equations for v_s and Q:

$$v_s = \frac{SCAP_V}{AP_A + SK} \tag{8.12}$$

$$Q = fv_s = \frac{fSCAP_V}{AP_A + SK}. \tag{8.13}$$

Note that these are algebraic equations which specify the values of v_s or Q corresponding to given values of the variables on the right which are valid for all times. A block diagram for Eq. (8.13) appears in Fig. 94. It is convenient to regard this isolated ventricle as having two different forms of input—a direct forcing P_V and a parametric forcing P_A. If we multiply the direct input P_V by the block transfer function, we get the output Q

but, unfortunately, the value of the transfer function depends on the parametric input P_A. This, as usual, means a nonlinearity.

We note in passing that one can use more complex models for both the filling and emptying processes. Thus filling models incorporating a time-dependent compliance have been explored.

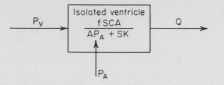

FIG. 94. BLOCK DIAGRAM OF ISOLATED VENTRICLE

While they successfully account for the initial phase of filling where intraventricular pressure is falling, the over-all performance is very close to that of the simpler model. Some preliminary explorations of an emptying process based on the Hill force-velocity relation and an assumed cylindrical ventricular geometry have also been made. Again, however, such a model appears to obey Starling's law as defined by Eq. (8.6) over the physiological range of load pressures. We will therefore be content with our simpler model for present purposes. Let us now turn to the circuits.

The isolated circuits. Our problem here is the converse of that just considered. There we defined the dependence of ventricular output on circuit pressures. Now we wish to define the dependence of circuit pressures on ventricular outputs. The appropriate experimental prototype is the cardiac by-pass preparation in which the pressure-dependent ventricles are replaced by mechanical pumps whose arbitrary outputs are independent of circuit pressures. For the circuit components we shall use the simplified model already described in connection with Fig. 92. Each circuit, pulmonary and systemic, will be considered to consist of two pure linear volume compliances (artery C_A and vein C_V) connected peripherally by a pure linear (Poiseuille) resistance R. One mechanical pump (the left) transfers blood at rate Q_L from pulmonary vein to systemic artery, and a second (the right) moves blood at rate Q_R from systemic vein to pulmonary artery. A hydraulic analog of this system appears in Fig. 95. The subscripts A and V in the figure refer to artery and vein, respectively, while P and S refer to pulmonary or systemic circuit. The system contains a fixed volume of blood B distributed

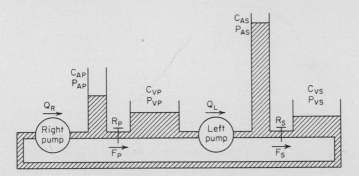

FIG. 95. HYDRAULIC ANALOG OF ISOLATED CIRCUITS

among its four reservoirs, the pumps and connecting tubing being assumed to represent a negligibly small volume. It is clear that what we have here is not one but four of Otto Frank's Windkessels connected in appropriate fashion.

It is quite easy to write a set of equations defining the operation of this system. We first formulate a continuity equation for each of the four reservoirs. Thus, the rate of change of blood volume in the systemic artery, $\dot{B}_{AS}$, is simply the difference between the rate of inflow Q_L and outflow F_S. But since the arterial compliance C_{AS} is by definition the ratio of volume to pressure (i.e., $C_{AS} = B_{AS}/P_{AS}$), it is apparent that $B_{AS} = C_{AS}P_{AS}$ and that $\dot{B}_{AS} = C_{AS}\dot{P}_{AS}$. We therefore have:

$$C_{AS}\dot{P}_{AS} = Q_L - F_S. \tag{8.14}$$

Analogous equations for each of the other three reservoirs can now be written at once:

$$C_{VS}\dot{P}_{VS} = -Q_R + F_S \tag{8.15}$$
$$C_{AP}\dot{P}_{AP} = Q_R - F_P \tag{8.16}$$
$$C_{VP}\dot{P}_{VP} = -Q_L + F_P. \tag{8.17}$$

Next we note that by Poiseuille's law:

$$F_S = \frac{P_{AS} - P_{VS}}{R_S} \tag{8.18}$$

and

$$F_P = \frac{P_{AP} - P_{VP}}{R_P} \tag{8.19}$$

and substituting in (8.14)–(8.17) and rearranging we have finally:

$$R_S C_{AS} \dot{P}_{AS} + (P_{AS} - P_{VS}) = R_S Q_L \tag{8.20}$$

$$R_S C_{VS} \dot{P}_{VS} - (P_{AS} - P_{VS}) = -R_S Q_R \tag{8.21}$$

$$R_P C_{AP} \dot{P}_{AP} + (P_{AP} - P_{VP}) = R_P Q_R \tag{8.22}$$

$$R_P C_{VP} \dot{P}_{VP} - (P_{AP} - P_{VP}) = -R_P Q_L. \tag{8.23}$$

This set of four simultaneous first-order linear differential equations defines the response of the four circuit pressures to direct forcing by Q_L and Q_R. It is not difficult to combine them to obtain equations in any one of the dependent pressure variables, just as we combined our respiratory equations (7.22) and (7.23) to obtain equations in a single dependent variable, θ_T or θ_A. Thus, if we solve (8.20) for P_{VS}, differentiate the result to obtain $\dot{P}_{VS}$, substitute these values into (8.21) and integrate the result, we obtain an equation in P_{AS} only. Similar manipulations yield equations for P_{VS}, P_{AP}, and P_{VP}:

$$\left(\frac{R_S C_{AS} C_{VS}}{C_{AS} + C_{VS}} \right) \dot{P}_{AS} + P_{AS}$$

$$= \frac{1}{C_{AS} + C_{VS}} \int (Q_L - Q_R) \, dt + P_0 + \frac{R_S C_{VS}}{C_{AS} + C_{VS}} Q_L \tag{8.24}$$

$$\left(\frac{R_S C_{AS} C_{VS}}{C_{AS} + C_{VS}} \right) \dot{P}_{VS} + P_{VS}$$

$$= \frac{1}{C_{AS} + C_{VS}} \int (Q_L - Q_R) \, dt + P_0 - \frac{R_S C_{AS}}{C_{AS} + C_{VS}} Q_R \tag{8.25}$$

$$\left(\frac{R_P C_{AP} C_{VP}}{C_{AP} + C_{VP}} \right) \dot{P}_{AP} + P_{AP}$$

$$= \frac{-1}{C_{AP} + C_{VP}} \int (Q_L - Q_R) \, dt + P_0 + \frac{R_P C_{VP}}{C_{AP} + C_{VP}} Q_R \tag{8.26}$$

$$\left(\frac{R_P C_{AP} C_{VP}}{C_{AP} + C_{VP}}\right) \dot{P}_{VP} + P_{VP}$$

$$= \frac{-1}{C_{AP} + C_{VP}} \int (Q_L - Q_R)\, dt + P_0 - \frac{R_P C_{AP}}{C_{AP} + C_{VP}}\, Q_L. \quad (8.27)$$

Let us examine some implications of these equations. First we may inquire as to the origin and significance of the term P_0. It represents a constant of integration which we have chosen to evaluate under initial conditions which are both convenient and physiologically meaningful. What we have assumed is that when $t = 0$, the total blood volume is so distributed that $P_{AS} = P_{VS} = P_{AP} = P_{VP} \equiv P_0$. This is the steady-state distribution which would result if $Q_L = Q_R = 0$ and the pumps were by-passed to allow the passive flow necessary to establish static-pressure equilibrium between the pulmonary and systemic circuits. Although our equations do not provide for such by-passes, the heart does. Thus P_0 is the static blood pressure or mean circulatory filling pressure which can be determined by producing cardiac arrest in an experimental animal. It is evidently equal to $[B/(C_{AS} + C_{VS} + C_{AP} + C_{VP})]$. Turning next to the integral term $\int (Q_L - Q_R)\, dt$, it is clear that this accounts for any net transfer of blood between pulmonary and systemic circuits occurring after $t = 0$. Again we note that once such transfer has occurred, our equations do not allow us to return to P_0 simply by setting $Q_L = Q_R = 0$, since no provision is made for passive flow between circuits.

Suppose now that starting from the initial condition, $P_{AS} = P_{VS} = P_{AP} = P_{VP} \equiv P_0$, Q_L and Q_R were simultaneously stepped from zero to Q_F. We would like to determine the response of the four pressures. For the prescribed form of forcing, the function $\int (Q_L - Q_R)\, dt$ is zero for all values of t, and so can be dropped. This leaves us with a familiar problem, i.e., that of obtaining particular solutions of linear first-order differential equations for step-function forcing. We can write the solutions at once without going through the details of the classical method, recalling that the term P_0 represents a bias or scale-shifting operation (Chapter 7):

$$P_{AS} = P_0 + \frac{R_S C_{VS}}{C_S} Q_F(1 - e^{-t/\tau_s}) \qquad (8.28)$$

$$P_{VS} = P_0 - \frac{R_S C_{AS}}{C_S} Q_F(1 - e^{-t/\tau_s}) \qquad (8.29)$$

$$P_{AP} = P_0 + \frac{R_P C_{VP}}{C_P} Q_F(1 - e^{-t/\tau_p}) \qquad (8.30)$$

$$P_{VP} = P_0 - \frac{R_P C_{AP}}{C_P} Q_F(1 - e^{-t/\tau_p}), \qquad (8.31)$$

where $C_S \equiv C_{AS} + C_{VS}$, $C_P \equiv C_{AP} + C_{VP}$,
$\tau_S = R_S C_{AS} C_{VS}/(C_{AS} + C_{VS})$, and $\tau_P \equiv R_P C_{AP} C_{VP}/(C_{AP} + C_{VP})$.

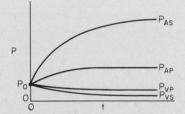

FIG. 96. TRANSIENT RESPONSE OF CIRCUIT PRESSURES TO
STEP-FUNCTION FORCING OF $Q_L = Q_R \equiv Q_F$

These solutions are plotted in Fig. 96. As $t \to \infty$, the exponential term drops out and the steady-state values of the four output pressures are seen to be linear functions of Q_F according to the following equations:

$$P_{AS} = P_0 + \frac{R_S C_{VS}}{C_S} Q_F \qquad (8.32)$$

$$P_{VS} = P_0 - \frac{R_S C_{AS}}{C_S} Q_F \qquad (8.33)$$

$$P_{AP} = P_0 + \frac{R_P C_{VP}}{C_P} Q_F \qquad (8.34)$$

$$P_{VP} = P_0 - \frac{R_P C_{AP}}{C_P} Q_F. \qquad (8.35)$$

These are plotted in Fig. 97.

On the other hand, if Q_L and Q_R went from zero to Q_F in such a way that $\int (Q_L - Q_R)\, dt \neq 0$ over the interval, then the steady-state values of the four pressures would be those given by Eqs. (8.32)–(8.35) plus or minus the equivalent static-pressure change within the appropriate circuit resulting from the shift of blood volume between systemic and pulmonary circuits. The transient behavior would depend upon the exact form

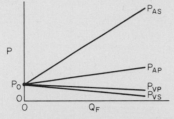

FIG. 97. STEADY-STATE DEPENDENCE OF CIRCUIT PRESSURES
ON PUMP OUTPUT

of $Q_L(t)$ and $Q_R(t)$. For example, with $P_{AS} = P_{VS} = P_{AP} = P_{VP} = P_0$ and $Q_L = Q_R = 0$ at $t = 0$, what would happen if we stepped Q_L to Q_F and then, at $t = t_1$, stepped Q_R to Q_F? In the interval $0 < t \leq t_1$, $(Q_L - Q_R) = Q_F$, and $\int (Q_L - Q_R)\, dt = Q_F t$, a ramp function. Hence the forcing in Eq. (8.24) during this interval is the sum of a ramp function and a step function. Recalling the superposition principle, the response to the two forcings applied simultaneously is the same as the sum of the response to each applied separately. The solution of (8.24) under these conditions is readily obtained by the classical method:

$$P_{AS} = P_0 + \frac{R_S C_{VS}}{C_S} Q_F(1 - e^{-t/\tau_s}) + \frac{Q_F}{C_S} t + \frac{\tau_S Q_F}{C_S} (e^{-t/\tau_s} - 1)$$

$$0 \leq t \leq t_1. \quad (8.36)$$

The first term is the familiar bias term, the second arises from the step function, and the last two from the ramp function. This latter contribution to the total response is plotted in Fig. 98, and when this is added to the first two terms, the total response is obtained. Now, when $t = t_1$, we step Q_R to Q_F and thereafter $(Q_L - Q_R) = 0$. Hence for $t > t_1$, the forcing term is a constant and the value of P_{AS} at the start of the interval is given by (8.36) with $t = t_1$. Thus the solution over this interval is simply a step-function response which is already familiar.

In a sense our system is incomplete, for it does not include a disturbing or noise signal. The simplest type of disturbance from the mathematical point of view happens also to be one

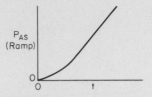

P_{AS}
(Ramp)

O

O   t

FIG. 98. RAMP FORCING CONTRIBUTION TO TOTAL RESPONSE DEFINED BY EQ. (8.36)

which is of considerable physiological interest, i.e., hemorrhage. Let us therefore introduce the quantity Q_H to represent the rate of hemorrhage, i.e., the rate of flow of blood from the system to the external environment. To include this term in our equations, we need to specify which of our four blood vessels is the source of the hemorrhage. We can choose any one (or more) we wish, but let us use the systemic artery as our example. We need only modify one equation of our original set, i.e., Eq. (8.14), which now becomes:

$$C_{AS}\dot{P}_{AS} = Q_L - F_S - Q_H. \tag{8.37}$$

It is easy enough to show that this will result in the following modifications of Eqs. (8.24) and (8.25):

$$\left(\frac{R_S C_{AS} C_{VS}}{C_{AS} + C_{VS}}\right)\dot{P}_{AS} + P_{AS} = \frac{1}{C_{AS} + C_{VS}} \int (Q_L - Q_R - Q_H)\, dt$$

$$+ P_0 + \frac{R_S C_{VS}}{C_{AS} + C_{VS}}(Q_L - Q_H) \tag{8.38}$$

$$\left(\frac{R_S C_{AS} C_{VS}}{C_{AS} + C_{VS}}\right)\dot{P}_{VS} + P_{VS} = \frac{1}{C_{AS} + C_{VS}} \int (Q_L - Q_R - Q_H)\, dt$$

$$+ P_0 - \frac{R_S C_{AS}}{C_{AS} + C_{VS}} Q_R. \tag{8.39}$$

Equations (8.26) and (8.27) remain the same as before. It is now apparent that if Q_H were stepped from zero to Q_{H_1}, with the system in some dynamic steady state with $Q_L = Q_R = Q_F$, the response of P_{AS} would be the sum of its response to a ramp and a step similar, except for sign, to that just described in the paragraph above.

Let us now obtain the normal Laplace transforms for Eqs. (8.38)–(8.39) and (8.26)–(8.27). We will omit the bias term P_0 and use our previously defined consolidated constants with the additional definition $\Delta Q \equiv (Q_L - Q_R)$. The results are*:

$$(\tau_{SS} + 1)P_{AS}(s)_N = \left[\frac{1}{C_{SS}}\right][\Delta Q(s) - Q_H(s)]$$
$$+ \frac{R_S C_{VS}}{C_S}[Q_L(s) - Q_H(s)] \quad (8.40)$$

$$(\tau_{SS} + 1)P_{VS}(s)_N = \left[\frac{1}{C_{SS}}\right][\Delta Q(s) - Q_H(s)] - \frac{R_S C_{AS}}{C_S}Q_R(s) \quad (8.41)$$

$$(\tau_{PS} + 1)P_{AP}(s)_N = \left[\frac{-1}{C_{PS}}\right]\Delta Q(s) + \frac{R_P C_{VP}}{C_P}Q_R(s) \quad (8.42)$$

$$(\tau_{PS} + 1)P_{VP}(s)_N = \left[\frac{-1}{C_{PS}}\right]\Delta Q(s) - \frac{R_P C_{AP}}{C_P}Q_L(s) \quad (8.43)$$

and rearranging into transfer-function form, we obtain:

$$P_{AS}(s)_N = \left[\frac{1}{C_{SS}(\tau_{SS} + 1)}\right][\Delta Q(s) - Q_H(s)]$$
$$+ \left[\frac{R_S C_{VS}}{C_S(\tau_{SS} + 1)}\right][Q_L(s) - Q_H(s)] \quad (8.44)$$

$$P_{VS}(s)_N = \left[\frac{1}{C_{SS}(\tau_{SS} + 1)}\right][\Delta Q(s) - Q_H(s)]$$
$$- \left[\frac{R_S C_{AS}}{C_S(\tau_{SS} + 1)}\right]Q_R(s) \quad (8.45)$$

$$P_{AP}(s)_N = \left[\frac{-1}{C_{PS}(\tau_{PS} + 1)}\right]\Delta Q(s) + \left[\frac{R_P C_{VP}}{C_P(\tau_{PS} + 1)}\right]Q_R(s)$$
$$(8.46)$$

$$P_{VP}(s)_N = \left[\frac{-1}{C_{PS}(\tau_{PS} + 1)}\right]\Delta Q(s) - \left[\frac{R_P C_{AP}}{C_P(\tau_{PS} + 1)}\right]Q_L(s).$$
$$(8.47)$$

* The subscript N refers to the normal response.

Equations (8.44)–(8.47) provide the basis for the block diagram of the isolated circuits shown in Fig. 99. In it we have added the bias value P_0 to the normal outputs in accordance with the practice of previous chapters. As usual, the Laplace transform-block diagram representation provides a very con-

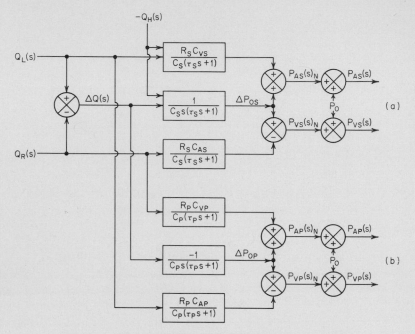

FIG. 99. BLOCK DIAGRAM OF ISOLATED CIRCUITS
(a) SYSTEMIC CIRCUIT; (b) PULMONARY CIRCUIT

venient and illuminating way of looking at the system. It emphasizes the fact that the only coupling between the systemic and pulmonary circuits is provided by the ΔQ input operating through the middle block of each circuit. Thus, if $\Delta Q = 0$, the two circuits operate completely independently of each other. It also makes clear that as long as the system is forced by the pump outputs, Q_L and/or Q_R, or by hemorrhage or transfusion, $\pm Q_H$, it behaves as a simple, directly forced linear system to which we can apply all of the mathematical machinery of previous chapters.

For example, it is very easy to synthesize the frequency response of the various pressures to sinusoidal forcing of Q_L,

Q_R, or Q_H. Let us provide a simple illustration of this by keeping $Q_L = Q_R$, letting $Q_H = Q_{H_1} \sin \omega t$, and looking at the behavior of P_{AS}. Since the frequency response of the latter is independent of its mean value (or *DC* component), we may just as well make it easy for ourselves by taking $Q_L = Q_R = P_0 = 0.$* The system under study can then be represented by the block diagram of Fig. 100, where $K_1 \equiv R_S C_{VS}/C_S$ and $K_2 \equiv 1/C_S$. Recalling our

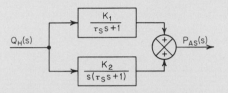

FIG. 100. SIMPLIFIED BLOCK DIAGRAM FOR Q_H FORCING OF SYSTEMIC ARTERIAL SYSTEM

discussion in Chapter 3, the transfer function of this parallel combination is the sum of the two transfer functions, i.e.,

$$P_{AS}(s) = \left[\frac{K_1}{\tau_S s + 1} + \frac{K_2}{s(\tau_S s + 1)}\right] Q_H(s) \qquad (8.48)$$

and performing the indicated algebraic addition:

$$P_{AS} = \left[\frac{K_1 s + K_2}{s(\tau_S s + 1)}\right] Q_H(s) = \left[\frac{K_2[(K_1/K_2)s + 1]}{s(\tau_S s + 1)}\right] Q_H(s)$$

$$= \left[\frac{K_2(\tau_1 s + 1)}{s(\tau_S s + 1)}\right] Q_H(s), \qquad (8.49)$$

where $\tau_1 \equiv K_1/K_2$.

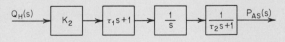

FIG. 101. CASCADE SYSTEM EQUIVALENT TO FIG. 100

It is now apparent that the over-all transfer function can be represented by the equivalent cascade system of Fig. 101, where all of the blocks are familiar ones. To synthesize the frequency response of P_{AS}, we need only add the gain and phase curves of

* It is also convenient to redefine Q_H as the rate of transfusion.

these four blocks on a Bode diagram. Figures 102 and 103 illustrate this process for certain arbitrary values of the system constants. It is apparent that even in this simple system considerable variation in shape of the composite curves may occur, depending upon the relative values of τ_1 and τ_S. Thus, if

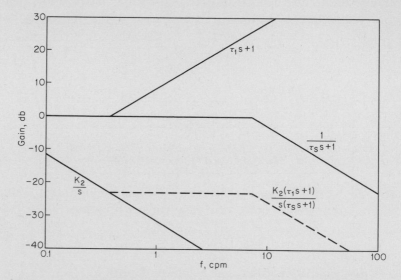

FIG. 102. APPROXIMATE (ASYMPTOTIC) BODE GAIN CURVES
FOR SYSTEM OF FIG. 101

$\tau_S/\tau_1 = 1$, the gain curve is a straight line of slope -6 db/octave crossing the zero db ordinate at $f = K_2/2\pi$. If $\tau_S/\tau_1 < 1$ as is true in Fig. 102, the gain curve will be flat between two break frequencies and have a slope of -6 db/octave on either side. The important system parameter which determines the ratio of these two time constants is the ratio of venous to arterial compliance, C_{VS}/C_{AS}.

However, the block diagram of Fig. 99 also identifies the nature of certain complications which we can expect as soon as we allow this system to be operated upon by neural and chemical signals. We have already noted that these signals operate upon R_S so that once again we encounter parametric forcing which will introduce additional nonlinearity into the closed-loop regulator. Before worrying about this, however, let us first

combine the isolated circuits with the two ventricles in order
to synthesize and examine the complete mechanical cardiovascular system. We will find that this process introduces complications of its own!

The complete mechanical system. If we now replace our mechan-

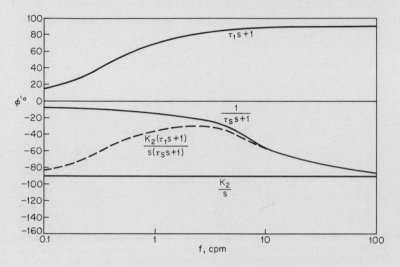

FIG. 103. BODE PHASE CURVES FOR SYSTEM OF FIG. 101

ical pumps with a right and left ventricle, Q_L and Q_R in Eqs.
(8.38)–(8.39) and (8.26)–(8.27) are no longer arbitrary but
instead become functions of circuit pressures. This dependence
on pressure was defined by Eq. (8.13) which we repeat here
with appropriate subscripts for each ventricle:

$$Q_L = \frac{f S_L C_L A_L P_{VP}}{A_L P_{AS} + S_L K_L} \tag{8.50}$$

$$Q_R = \frac{f S_R C_R A_R P_{VS}}{A_R P_{AP} + S_R K_R}. \tag{8.51}$$

Simultaneous solution of the set of equations, (8.50)–(8.51),
(8.38)–(8.39), and (8.26)–(8.27) will define the behavior of this
combined system whose block diagram appears in Fig. 104. It
is apparent that the combined equations will be nonlinear because of the form of (8.50)–(8.51). Before attacking the problem

of solving these equations, let us first decide what sort of questions we are going to ask this system to answer.

The first question we ourselves should try to answer is whether or not this system should be regarded as a regulator and, if so, what quantity or quantities it might regulate. It soon becomes apparent that there is more than one way of looking at this problem. For example, Fig. 104 appears as if it might represent

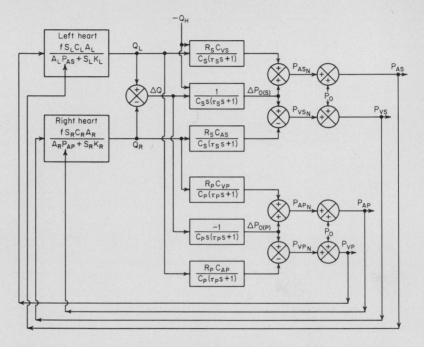

FIG. 104. BLOCK DIAGRAM OF COMBINED MECHANICAL SYSTEM

a feedback system designed to regulate all or some of the circuit pressures. From this viewpoint, the ventricles would play the role of the controlling system and the circuits would represent the controlled system. If we ask how this system responds to the parametric disturbance of a decreased R_S, we find that an increase in cardiac output occurs and that this does indeed reduce the error in P_{AS} which would result from a similar disturbance in the open-loop system of Fig. 99. However, it certainly does not help to regulate P_{AS} against the direct disturbance of hemorrhage, for here it turns out that a fall in cardiac output, which exaggerates the error, occurs. The same is true for the parametric disturbance of increased C_{VS}. Hence it seems clear that this system does not function primarily as a regulator of circuit pressures.

There is one quantity, however, which does appear to be consistently regulated by this system alone and which, moreover,

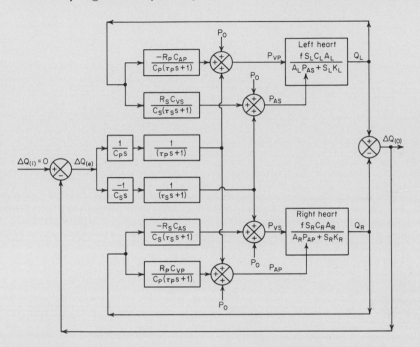

FIG. 105. ALTERNATIVE BLOCK DIAGRAM OF THE MECHANICAL
SYSTEM EMPHASIZING ΔQ REGULATION

is characterized by zero steady-state error. This quantity is ΔQ, and its control constitutes the mechanical self-regulation previously mentioned in this chapter. To bring out this feature of the system, let us redraw Fig. 104 in a new form (Fig. 105) in which it is apparent that the identification of controlling and controlled systems has been reversed. We also note the integral blocks in the controller which zero the steady-state error in ΔQ. The essence of the system can perhaps be brought out more clearly by omitting some of the details and linearizing it as in Fig. 106.

Here we have let P_L represent that combination of pressure changes which increases Q_L (i.e., $+P_{VP}$ and $-P_{AS}$), P_R the combination which increases Q_R (i.e., $+P_{VS}$ and $-P_{AP}$), and have lumped all of the controller dynamics except the integral action into the two process blocks. It is now apparent that in

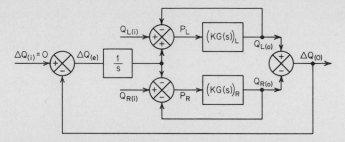

FIG. 106. SIMPLIFIED VERSION OF FIG. 105 ILLUSTRATING THE SEVERAL MECHANICAL CONTROL ACTIONS INVOLVED

the absence of the coupling provided by the ΔQ loop, we would have two independent proportional control systems regulating Q_{L_0} and Q_{R_0}, respectively. Adding the ΔQ loop makes the two systems interdependent, and the integral action zeros the steady-state error of ΔQ. This mechanical self-regulation is a very important stabilizing feature which automatically limits the duration of inequalities in Q_L and Q_R from any cause.

Although some cardiovascular physiologists might rebel at the notion of regarding the entire peripheral circulation together with the pressure-dependent properties of the ventricles as a control device to ensure equality of Q_L and Q_R, it is nevertheless clear that this is one of its important functions.[*] Moreover, elucidation of how the steady-state values of cardiac output and circuit pressures at which this equality is established depend upon heart and circuit properties has long been of major concern to physiologists and clinicians alike. It is therefore appropriate that this be the first question we put to our mechanical system.

Steady-state behavior. There are several possibilities for obtaining the steady-state solutions of our set of equations for particular

---

[*] W. F. Hamilton, "Role of Starling concept in regulation of normal circulation," *Physiol. Revs.* 35 (1955): 161.

values of the system properties. For example, we could derive algebraic equations specifically applicable to the steady state and solve these, and this has been done elsewhere.* A much easier and more flexible procedure is to program the dynamic equations on an analog computer and let the solutions run to the desired steady state. This is the procedure which we shall use here. Again, there are various possibilities for designing appropriate analog circuitry, and we shall base ours directly upon Eqs. (8.20)–(8.23) and (8.50)–(8.51). The corresponding computer diagrams are shown in Figs. 107 and 108, the latter

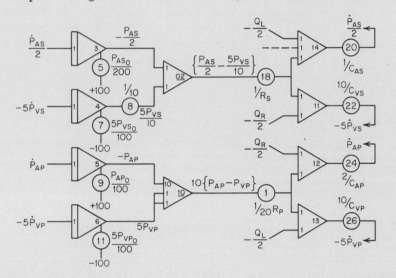

FIG. 107. ANALOG COMPUTER CIRCUITRY FOR SOLUTION OF CIRCUIT EQS. (8.20)–(8.23)

employing servodividers which we have not previously encountered. For the complete mechanical system, the pressure inputs of Fig. 108 are taken from the outputs of amplifiers 3, 4, 5, and 6 of Fig. 107, and the Q_L and Q_R outputs of Fig. 108 are applied to amplifiers 11, 12, 13, and 14 of Fig. 107. To obtain steady-state solutions for any given set of system properties it is only necessary to set the coefficient potentiometers to the appropriate

* F. S. Grodins, "Integrative cardiovascular physiology: a mathematical synthesis of cardiac and blood vessel hemodynamics," *Quart. Rev. Biol.* 34 (1959): 93.

values and choose initial conditions for P_{AS}, P_{VS}, P_{AP}, and P_{VP} subject to the following restriction:

$$C_{AS}P_{AS_0} + C_{VS}P_{VS} + C_{AP}P_{AP_0} + C_{VP}P_{VP_0} = B. \quad (8.52)$$

Integration is then allowed to proceed until a steady state is reached. If the initial conditions chosen differ from the steady-

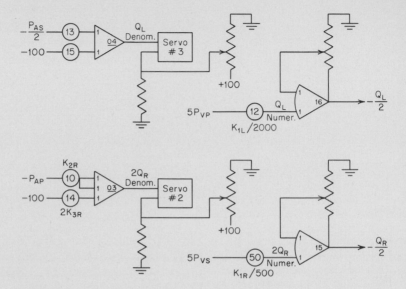

FIG. 108. ANALOG COMPUTER CIRCUITRY FOR SOLUTION OF HEART EQS. (8.50)–(8.51)

state values, as will generally be the case, the dynamic portion of the response corresponds to free recovery from an initial displacement. Let us examine some results of this sort of exploration.

Figure 109 illustrates the dependence of steady-state cardiac output and circuit pressures upon blood volume. All other system parameters were assigned values estimated as normal for an adult human. This is perhaps the simplest form of forcing and the responses are those which would be expected intuitively. It is apparent that this isolated mechanical system has no effective way of maintaining Q and P_{AS} in the face of hemorrhage.

Figure 110 shows the steady-state dependence of cardiac output and circuit pressures upon arteriolar resistance in the pulmonary and systemic circuits. There are two points worth noting. One is the considerable difference between the effects of R_P and R_S, and this is critically dependent upon the values of the system parameters. The other is the fact that when R_S falls, Q rises. We recall from our earlier discussion that when

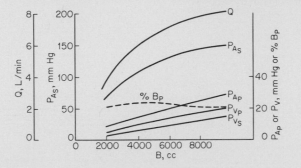

FIG. 109. STEADY-STATE DEPENDENCE OF CARDIAC OUTPUT AND CIRCUIT PRESSURES ON TOTAL BLOOD VOLUME

B_P = Pulmonary blood volume. From F. S. Grodins, "Integrative cardiovascular physiology," *Quart. Rev. Biol.* 34 (1959): 93.

tissue metabolism increases, R_S falls as a result of the operation of the chemical loops of Fig. 91. Is this a way of adjusting cardiac output to tissue metabolism without the need for any other neural or humoral mechanisms, and could this be responsible

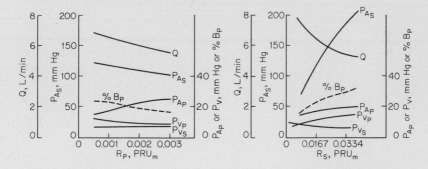

FIG. 110. STEADY-STATE DEPENDENCE OF CARDIAC OUTPUT AND CIRCUIT PRESSURES ON PULMONARY (LEFT) AND SYSTEMIC (RIGHT) ARTERIOLAR RESISTANCE

From F. S. Grodins, "Integrative cardiovascular physiology," *Quart. Rev. Biol.* 34 (1959): 93.

for the increase in cardiac output during exercise? The answer to the first question is yes but to the second is no. The maximum increase in Q produced by this means alone is much less than that observed in exercise, and moreover, it is accompanied by a fall in P_{AS}. It turns out that in order for this system to reproduce the observed response to exercise, it is necessary to alter both heart and circuit properties in the manner summarized in Table 6.

TABLE 6. RESPONSES AND FORCINGS IN EXERCISE[a]

Responses	Rest	Exercise	Forcings	Rest	Exercise
Q, L/min.	6.5	27	f, cpm	75	150
v_s, cc	86.6	180	S_L, mm Hg	50	135
$(w_s)_L$, g m	137	368	S_R, mm Hg	7	22.5
$(w_s)_R$, g m	20.8	61.5	$R_L C_L$, sec	0.35	0.09
$(v_d)_L$, cc	201	200	$R_R C_R$,[b] sec	0.35	0.28
$(v_d)_R$, cc	216	200	R_S, mm Hg/cc/min.	0.0167	0.0053
$(v_r)_L$, cc	114	20	R_P, mm Hg/cc/min.	0.001	0.00052
$(v_r)_R$, cc	129	20	C_{VS}, cc/mm Hg	500	490
P_{AS}, mm Hg	116	150			
P_{AP}, mm Hg	17.5	25			
P_{VS}, mm Hg	7.5	7.5			
P_{VP}, mm Hg	11.0	11.0			

[a] From F. S. Grodins, "Integrative cardiovascular physiology," *Quart. Rev. Biol.* 34 (1959): 93.

[b] C_L and C_R constant.

Finally we have illustrated the responses to changes in ventricular strength in Fig. 111. This has always been of major clinical interest for these are the uncomplicated changes resulting from a pure "hemodynamic sequence"* in left and in right heart failure. It seems obvious that this mechanism, though not sufficient, is probably a necessary link in the chain of events leading to the classical clinical picture of chronic congestive heart failure.

Dynamic behavior. We shall only present a few examples of the dynamic behavior of our mechanical system as obtained from the analog computer circuit of Figs. 107 and 108. Free recovery of the system from non-steady-state initial conditions is shown in Fig. 112. The final values attained correspond to the steady state for this particular set of system properties, and this is the

* Homer Smith, *The Kidney* (New York, Oxford University Press, 1951).

way in which the steady-state behavior described in the previous paragraph was established. Figure 113 shows the response to a 1.0-cpm Q_H square wave applied to the systemic artery. At $t = 10$, an arterial transfusion at the constant rate of 20 cc/sec was instantaneously established, and at $t = 40$, it was abruptly

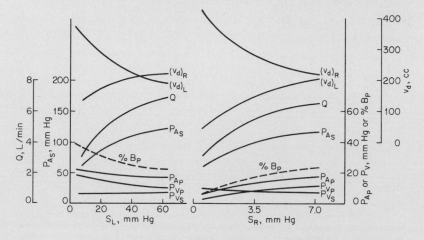

FIG. 111. STEADY-STATE DEPENDENCE OF CARDIAC OUTPUT, CIRCUIT PRESSURES, AND VENTRICULAR DIASTOLIC VOLUMES ON LEFT AND RIGHT VENTRICULAR STRENGTH

From F. S. Grodins, "Integrative cardiovascular physiology," *Quart. Rev. Biol.* 34 (1959): 93.

converted into a 20-cc/sec hemorrhage. Hemorrhage and transfusion then continued to alternate at 30-sec intervals. When the rate of hemorrhage and transfusion is increased to 30 cc/sec and the duration of each phase reduced to 7.5 sec (i.e., $f = $ 4 cpm), the system behaves as shown in Fig. 114. Finally, Fig. 115 shows the response to the sinusoid, $Q_H = Q_{H_0} \sin \omega_f t$, with $Q_{H_0} = 100$ cc/sec and $\omega_f/2\pi = 4$ cpm applied to the systemic artery. Despite the known nonlinearity of the system, the output wave forms are very nearly sinusoidal.

Synthesis: the cardiovascular chemostat

How shall we fit the model of the mechanical cardiovascular system just developed into the cardiovascular chemostat of

Fig. 91? Recall first that we developed this model because it was regarded as important for explaining just how the function $Q = \phi(P_{AS_{(a)}})$ in Fig. 91 depended upon the mechanical parameters of the heart and circulation. Presumably, then, the equations and block diagrams just obtained should answer this question and we need merely substitute this detailed answer for the black box of Fig. 91. However, this optimistic outlook is something of an illusion, for we have really only shifted our

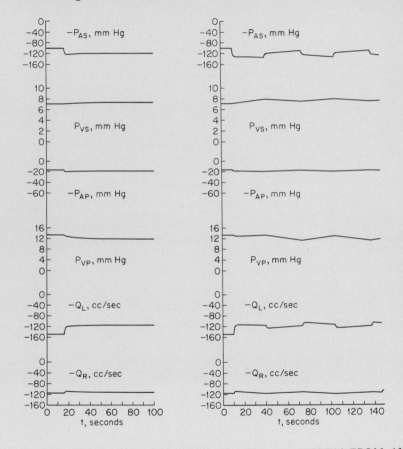

FIG. 112 (LEFT). FREE RECOVERY OF DYNAMIC SYSTEM FROM AN INITIAL DISPLACEMENT

FIG. 113 (RIGHT). DYNAMIC RESPONSE TO ARTERIAL Q_H SQUARE WAVE
Amplitude, 20 cc/sec; frequency, 1 cpm.

ignorance one link further back along the causal chain. Although we now know what will happen to Q if certain cardiac and/or circuit parameters are modified, we still need detailed information on just how these parameters are affected by changes in $P_{AS(a)}$. Such information is very incomplete and it is apparent from the outset that our attempts at synthesis can represent at best only an initial formulation of the problem and a preliminary guide for future explorations.

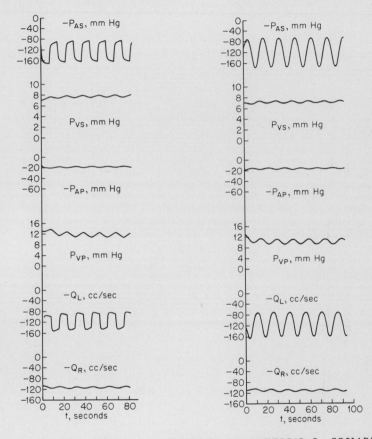

FIG. 114 (LEFT). DYNAMIC RESPONSE TO ARTERIAL Q_H SQUARE WAVE
Amplitude, 30 cc/sec; frequency, 4 cpm.

FIG. 115 (RIGHT). RESPONSE TO ARTERIAL Q_H SINUSOID
Amplitude, 100 cc/sec; frequency, 4 cpm.

A preliminary model: the arterial pressostat. Let us therefore begin very modestly with a model which omits the direct chemical loops of Fig. 91 and once more lumps all parallel systemic components into one. We shall further assume that the only parameters operated upon by $P_{AS_{(a)}}$ are R_S and f. Now we really do not know what analytical forms to assign to these functions, either on a theoretical or empirical basis. However, since our present purpose is to point out a general procedure rather than to obtain specific results, let us make as reasonable guesses as we can and see where they take us.

There is some reason to believe that arteriolar smooth muscle responds approximately as a first-order lag with a time constant on the order of 5–10 sec. Accordingly, let us write:

$$T_R \dot{R}_S + R_S = \phi_R(P_{AS_{(a)}}), \qquad (8.53)$$

where $\phi_R(P_{AS_{(a)}})$ is the steady-state gain function shown in Fig. 116. The latter was derived from the Blutdruck-Charakter-

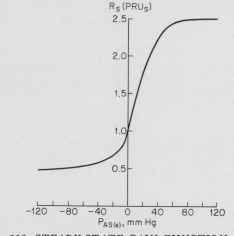

FIG. 116. STEADY-STATE GAIN FUNCTION, $\phi_R(P_{AS_{(a)}})$

istik curves of E. B. Koch[*] by substituting R_S for P_{AS_o} on the ordinate. This substitution implies the assumption that Q remains constant and this is almost certainly untrue. There is also evidence that the heart rate responds almost instantaneously

---

[*] E. B. Koch, "Die reflektorische Selbststeurung des Kreislaufes," *Ergeb. Kreislaufforschung* 1 (1931): 1.

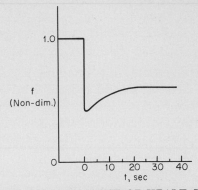

FIG. 117. TRANSIENT RESPONSE OF HEART RATE TO STEP
PRESSURE FORCING OF ISOLATED CAROTID SINUSES IN
THE DOG

to a step change in $P_{AS_{(e)}}$ and usually shows an initial undershoot (Fig. 117). The following equation will therefore be used
to describe this behavior:

$$\tau_f \dot{f} + f = \phi_f(P_{AS_{(e)}}) + k_f \dot{P}_{AS_{(e)}}, \qquad (8.54)$$

where $\phi_f(P_{AS_{(e)}})$ is the steady-state gain curve (again based on
Koch) shown in Fig. 118. In terms of the hardware involved,
we might imagine that τ_f represents a short first-order lag resid-

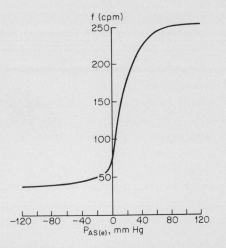

FIG. 118. STEADY-STATE GAIN FUNCTION, $\phi_f(P_{AS_{(e)}})$

ing in the sinus node and that the term $k_f \dot{P}_{AS_{(e)}}$ expresses a rate sensitivity characterizing the carotid sinus baroceptors. The ratio k_f/τ_f determines the magnitude of the initial undershoot. Unfortunately, the on and off f transients observed in the dog are not always symmetrical, thus implying a nonlinearity which we have not included. It should therefore be abundantly clear that Eqs. (8.53) and (8.54) are intended only as rough first approximations and are not to be taken too seriously. Moreover, it is very probable that other parameters (e.g., S_L, S_R, C_{VS}) also respond to $P_{AS_{(e)}}$.

A block diagram of the closed-loop system combining these controlling operations with the mechanical controlled system of Fig. 104 is shown in Fig. 119. Although the dynamic transfer functions of Eqs. (8.53) and (8.54) have all been placed in a single controlling system block for convenience, it should be clear that they are probably located in a variety of anatomical sites. We can appropriately call this control system the systemic arterial pressostat, a regulator designed to keep P_{AS} near normal in the presence of certain disturbances. The particular disturbance shown in the diagram is the simplest one, i.e., hemorrhage, but there are obviously many other possibilities. Have we abandoned the concept of a cardiovascular chemostat? Not at all, for it is still implicit in Fig. 119 if we remember that the components of R_S operated upon by P_{AS_e} do not include those of brain and heart. We recall that the controlled system itself was nonlinear because of the hyperbolic heart functions, and now we see that the controlling operations introduce parametric forcing and thus additional nonlinearity.

In exploring the behavior of the isolated mechanical system, we did not bother to combine heart and circuit equations explicitly, for it was much more convenient to work with the original equations in designing an analog computer circuit for the solution of this nonlinear system. The same is true in even greater measure for the closed-loop system of Fig. 119. Accordingly, we will simply modify the computer circuitry of Figs. 107 and 108 to include Eqs. (8.53) and (8.54). It turns out that the term "simply" is relatively appropriate for the modification required to include the R_S loop of Eq. (8.53). We need only

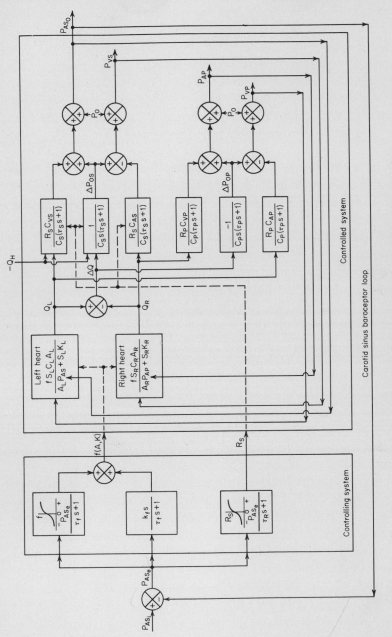

FIG. 119. BLOCK DIAGRAM OF ARTERIAL PRESSOSTAT

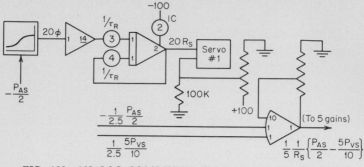

FIG. 120. *ANALOG COMPUTER CIRCUITRY FOR R_S LOOP*

replace amplifier (02) and potentiometer (18) in Fig. 107 by the circuit of Fig. 120. The left half of this circuit uses a function generator to form $\phi_R(P_{AS_{(o)}})$ and then solves the following rearranged form of Eq. (8.53):

$$\dot{R}_S = \frac{1}{\tau_R}\,[\phi_R(P_{AS_o}) - R_S]. \qquad (8.55)$$

The right half of the circuit uses a servodivider to form the term $(1/R_S)(P_{AS} - P_{VS})$. It is far from simple however to include the f loop of Eq. (8.54) in the computer circuit, for we recall that the A's and K's of Eqs. (8.12) and (8.13) are exponential

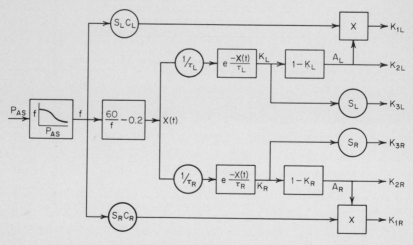

FIG. 121. *CALCULATION FLOW DIAGRAM FOR f LOOP*

functions of f [Eqs. (8.10) and (8.11)]. If we let $K_{1L} \equiv fS_LC_LA_L$, $K_{2L} \equiv A_L$, $K_{3L} \equiv S_LK_L$, $K_{1R} \equiv fS_RC_RA_R$, $K_{2R} \equiv A_R$, and $K_{3R} \equiv S_RK_R$, the calculations required to incorporate the f loop can be summarized in the diagram of Fig. 121. Appropriate computer circuitry can be designed to accomplish this* but we will not present it here.

We will only give a few illustrative examples of the behavior of this model both because its computer exploration and experimental testing have only just begun, and also because its equations will undoubtedly require extensive modification in the near future. Again, we are more interested in illustrating the general approach than in the specific results obtained at this stage. Let us look first at $P_{AS_{(o)}}$ forcing in the open-loop system. The corresponding animal preparation would be one in which both carotid sinuses were isolated as blind sacs† and in which all other arterial baroceptors were denervated. In Fig.

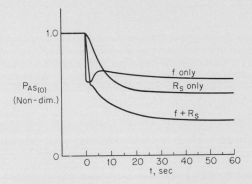

FIG. 122. THEORETICAL OPEN-LOOP RESPONSES TO STEP-FUNCTION PRESSURE FORCING UNDER DIFFERENT CONTROL CONDITIONS

122, we have illustrated the response of $P_{AS_{(o)}}$ to a step function $-P_{AS_{(o)}}$ forcing in the open-loop model when only the f, only the R_S, or both of these controlling operations are included.

The most significant feature illustrated by these curves is the

* William E. McAdam, Jr., An Analog Simulation of the Mammalian Cardiovascular System. Ms thesis, Dept. of Electrical Engineering, Northwestern University, 1961.

† E. Moissejeff, "Zur Kenntnis des Carotissinusreflexes," *Z. ges. exptl. Med.* 53 (1926): 696.

fact that a fast response is entirely dependent upon the f loop, i.e., upon a very prompt bradycardia. With the particular set of parameters used to generate these curves, the initial jog in the f loop response is almost entirely masked when it is combined with R_S control. For comparison, Fig. 123 illustrates the re-

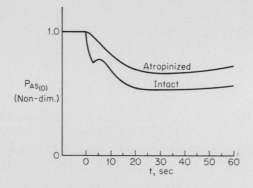

FIG. 123. OPEN-LOOP RESPONSES TO STEP-FUNCTION PRESSURE FORCING IN THE DOG

sponses observed in the dog when the motor vagi to the heart are intact and when they are blocked by atropine. The fast response in the intact animal is also due to a prompt bradycardia which undershoots somewhat to produce the jog in the response curve. When the bradycardia is blocked by atropine, the response is much slower. These features of prototype behavior thus agree reasonably well with the predictions of the model. However, there are obviously many details which remain to be included. Thus the pattern of bradycardia observed in the dog is quite variable and a type of vagal escape often occurs.* The output pressure frequently tends to rise gradually with time, and the off transient is generally different from the on. Finally, fast records show a dead-time (latent period) on the order of one second which has not been included in our model. These defects should become more apparent when we examine the response to other forms of the input function $P_{AS_{(o)}}(t)$.

Let us therefore look briefly at sinusoidal forcing of the open-

---

* S. C. Wang and H. L. Borison, "An analysis of the carotid sinus cardiovascular reflex mechanism," *Am. J. Physiol.* 150 (1947): 712.

loop system remembering, however, that many of the special advantages of frequency analysis cannot be realized in a non-linear system. The response of systemic arterial pressure to sinusoidal-pressure forcing of the isolated carotid sinuses of a dog at two frequencies is shown in Fig. 124. Perhaps the most

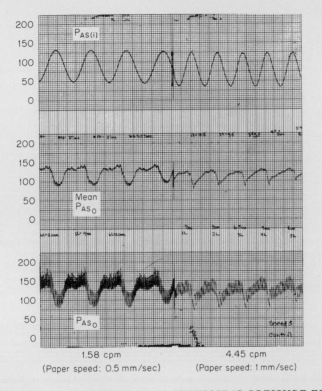

1.58 cpm   4.45 cpm
(Paper speed: 0.5 mm/sec)   (Paper speed: 1 mm/sec)

FIG. 124. OPEN-LOOP RESPONSE TO SINUSOIDAL PRESSURE FORCING IN THE DOG AT TWO DIFFERENT FREQUENCIES

striking thing about the response is that although periodic it is not a sine wave. As pointed out by Stegemann in 1957[*] there is a marked asymmetry, the pressure falling much faster than it rises. In addition, the output-pressure curve shows a peculiar twin peaking at the lower frequency. The response of the model

[*] J. Stegemann, "Der Einfluss sinusförmiger Druckänderung im isolierten Karotissinus auf Blutdruck und Pulsfrequenz beim Hund," *Deut. Ges. Kreislaufforsch.* 23 (1957): 392.

to these same two frequencies is shown in Fig. 125, the solid curve representing a relatively high, and the dashed curve a relatively low, R_S loop gain. The output of the model seems to show the proper type of asymmetry (i.e., the pressure falls more rapidly than it rises) and the wave form approaches that

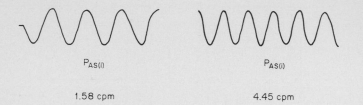

$P_{AS(i)}$ $P_{AS(i)}$

1.58 cpm   4.45 cpm

$P_{AS(0)}$ $P_{AS(0)}$

FIG. 125. THEORETICAL OPEN-LOOP RESPONSE TO SINUSOIDAL PRESSURE FORCING AT TWO FREQUENCIES (RETRACED FROM ORIGINAL RECORDS)

of the dog quite closely when the R_S gain is low. Although this is promising, it is also perhaps surprising, for we know we have omitted many details from our model. In a very complex system such as this one, it is quite possible to get the right answer for the wrong reason, and it will take much additional work to decide whether this is the case here. The curves in Fig. 125 were redrawn from the original computer records (in which $-P_{AS(0)}$ was recorded to economize on amplifiers) to allow easier comparison with the animal records in Fig. 124. An original computer record for Q_H sinusoidal forcing at 4.45 cpm with only the R_S loop, only the f loop, or both loops operating is shown in Fig. 126. Again, it is apparent that the asymmetry of the $P_{AS(0)}$ wave form is due mainly to the f loop.

It is therefore of interest that when the heart rate response is greatly diminished by vagotomy in the dog, the output wave form becomes very nearly sinusoidal (cf. R_S only in Fig. 126) and the system behaves essentially like a second-order linear one, at least over a restricted frequency range. Figure 127 is a

Bode diagram taken from such an experiment. The gain curve approaches an asymptotic slope of -12 db/octave at frequencies above about 1.5 cpm, and resembles that of a second-order system with a natural frequency of 1.0 cpm and damping ratio

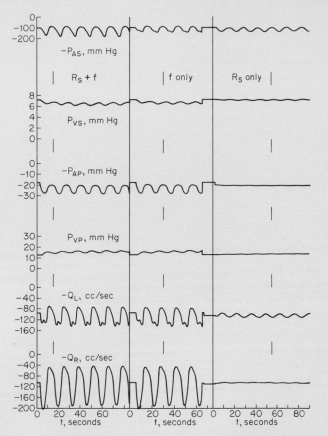

FIG. 126. COMPUTER OUTPUTS FOR SINUSOIDAL PRESSURE FORCING OF OPEN-LOOP SYSTEM AT 4.45 CPM WITH ONLY THE R_S, ONLY THE f, OR BOTH CONTROL LOOPS IN OPERATION

of 1.0. The phase curve taken directly from the data (solid line) shows a lag of $90°$ at the corner frequency (like a second-order system) but increases above $180°$ at higher frequencies (unlike a second-order system). If we assume the latter to be due to the presence of a dead time of 1.5 sec, we can correct for this by subtracting 9°/cpm* from the total phase curve, leaving the dashed curve to represent the basic second-order system. From the control standpoint we can thus conclude that the open-loop system is stable and that the closed-loop system should also be stable since we have a gain margin of some 12 db and a phase

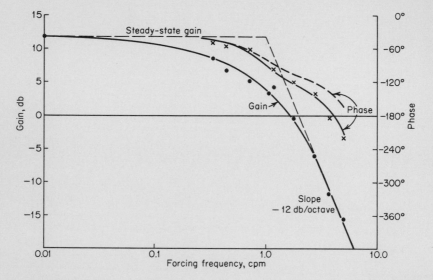

FIG. 127. BODE DIAGRAM OF OPEN-LOOP FREQUENCY RESPONSE OF A VAGOTOMIZED DOG

margin of about 60°. This conclusion is supported by the Nyquist *G*-plot in Fig. 128. The transfer locus in the *G*-plane passes to the right of the critical point $(-1, j0)$ thus implying a stable closed-loop system. It is obvious that any successful theoretical model of the arterial pressostat must account for this observed behavior. Unfortunately, we have not yet explored our model over a sufficient frequency range to know whether it does or not, nor do we yet have an adequate experimental comparison of the behavior of the intact and vagotomized dog.

Before leaving this model, let us briefly consider another sort of exploration which is currently in progress. We wish to look

* A dead time (transport lag, latent period) component does not affect the gain curve but adds a phase lag equal to $(360\ Tf)°$, where T is the dead time and f is the forcing frequency. Its transfer function is e^{-Ts}.

at the closed-loop system and ask how effectively it regulates against hemorrhage. To study this question we may apply a sinusoidal Q_H signal to the systemic artery in the model and in the dog. Figure 129 is an example of the output of the model at a forcing frequency of 4.45 cpm when only the R_S, only the f,

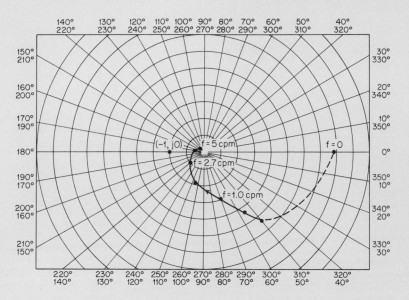

FIG. 128. NYQUIST G-PLANE PLOT OF OPEN-LOOP FREQUENCY RESPONSE OF A VAGOTOMIZED DOG

or both control loops are operating. It is apparent that control is considerably more effective when the f loop is operating because of the greater corrective contribution of cardiac output. Similar studies are in progress in the dog, but no detailed comparisons are yet available.

The cardiovascular chemostat. Let us finally conclude this chapter by returning to the cardiovascular chemostat described earlier and illustrated in Fig. 91. We would like to modify this figure to give a preview of the sort of model we eventually hope to be able to build. We can do little more than indicate its general nature here for many unknowns remain to be explored. Our tentative block diagram appears in Fig. 130. In it we have combined the mechanical cardiovascular system with both tissue and pulmonary gas exchangers and have also included all of the controlling operations discussed earlier. Note that the diagram of the mechanical circuit has been changed from that of Fig. 119 in order to retain F_{1-n} and F_P as explicit variables, as well as to isolate resistance and compliance elements. The

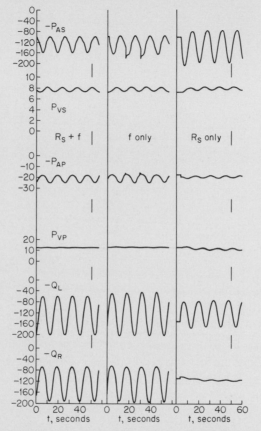

FIG. 129. THEORETICAL CLOSED-LOOP RESPONSE TO ARTERIAL Q_H SINUSOID AT 4.45 CPM UNDER DIFFERENT CONTROL CONDITIONS

transfer functions for this version are based directly upon Eqs. (8.14)–(8.19). Simplified exchanger transfer functions for O_2 only, which ignore the details of the blood-air phase difference in the lung, and the O_2 dissociation curve have been used. We can best describe this model as an adaptive control system

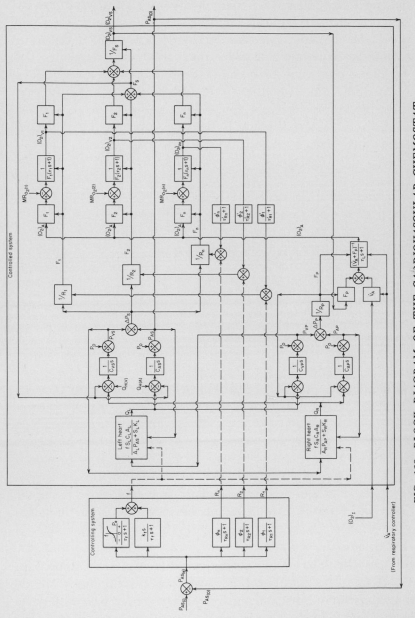

FIG. 130. BLOCK DIAGRAM OF THE CARDIOVASCULAR CHEMOSTAT

designed to regulate tissue (venous) oxygen levels. If we assign definite values to all of its transfer functions, solution of these equations with the aid of an analog or digital computer will reveal the dynamic responses of all the dependent variables to a variety of forcings.

Having presented this block diagram with only a very inadequate discussion, I shall say no more about it but simply leave it for the reader to study. I trust that he will marvel, as I do, at the multitude of beautifully balanced dynamic interrelationships which exist even in this grossly oversimplified model of a biological control system. When we know all of the transfer functions in Fig. 130 plus many others, some obviously and some not at all obviously omitted, then we will really begin to understand the mysteries of cardiovascular regulation. It is quite clear that what we have done in this chapter was to formulate rather than solve this problem.

Summary

The cardiovascular system is an extremely complex hydrodynamic system upon which is superimposed a variety of neural and humoral controlling operations. From the control standpoint, it appears to have many features in common with the respiratory chemostat, and it seems fruitful to regard it as an adaptive venous chemostat. For a variety of reasons, the chemostat function has not been widely emphasized. One reason is that control of venous composition is accomplished indirectly by regulating systemic arterial pressure, and this pressostat function is a much more obvious one. Another reason is that mechanical rather than chemical forcings have long been the major concern of cardiovascular physiologists and clinicians alike. Hence, unlike the respiratory chemostat, any model of cardiovascular regulation which by-passes the mechanical features of the heart and blood vessels cannot be regarded as satisfactory. We have therefore developed a simplified dynamic model of the mechanical cardiovascular system which revealed among other things that the system operates as an integral controller to ensure a zero steady-state error in ΔQ.

The model was then extended to include the baroceptor feedback loop. The equations of this pressostat were nonlinear both because of hyperbolic cardiac transfer functions, and because of feedback through the parameters f and R_S. Despite the known internal complexities of this pressostat, a conventional open-loop frequency analysis in the vagotomized dog revealed a behavior essentially like that of a critically damped second-order system, from which it was apparent that the closed-loop system would be stable. Finally, the model was extended to include pulmonary and tissue gas exchangers and explicitly incorporate the chemostat function. Some limited computer explorations have been made but much more remains to be done. We have really formulated rather than solved the problems of cardiovascular regulation considered in this chapter.

CHAPTER 9

In Retrospect and in Prospect

IT is obvious that we have only scratched the surface of the vast field of biological control systems. Many facets of respiratory and cardiovascular regulation remain to be explored both experimentally and theoretically before adequate understanding can be achieved. It is perhaps difficult to believe that despite many years of intensive investigation we do not yet have a satisfactory explanation for the most important physiological response of all, i.e., the response to exercise. Control-system theory alone cannot solve this problem, but it can serve as a valuable guide to fruitful experimentation and analysis. The physiologist necessarily has a somewhat ambivalent attitude in his approach to biological control systems. On the one hand he is ultimately interested in the performance of the over-all system, but on the other he wants to understand this performance in terms of the contributions of individual components. Thus, although the elegant dynamic analysis of the cerebral ischemic pressor response reported by Sagawa, Taylor, and Guyton[*] provides much useful information about over-all performance, the reader may be disappointed to find that the dependence of the empirical system constants upon particular physiological components has yet to be established. However, control-system theory can also help design experiments to answer this sort of question, and the recent work of Warner and Cox[†] on the neural control of heart rate is a good example.

[*] K. Sagawa, A. E. Taylor, and A. C. Guyton, "Dynamic performance and stability of cerebral ischemic pressor response," *Am. J. Physiol.* 201 (1961): 1164.
[†] H. R. Warner and A. Cox, "A mathematical model of heart rate control by sympathetic and vagus efferent information," *J. Appl. Physiol.* 17 (1962): 349.

Most biological control systems still await application of the conceptual approaches and analytical tools of the mathematician, physicist, and control engineer. In many cases our knowledge is inadequate to permit even a satisfactory formulation of the problem in quantitative terms. The basic difficulty was perhaps best stated by a physicist when he said that a physical problem is usually so simple that there is little trouble in recognizing the presence of an interpretative paradox, but that a biological problem is usually so complex that one is never quite sure whether such a paradox exists or not! Nevertheless, the trend seems clear and the methods of control-system engineering are being applied to an increasing variety of biological regulators.*

Despite this cheerful outlook, however, we cannot help but feel that something is missing. The first five chapters outlined the very powerful and satisfying analytical machinery which is available for the study of linear systems. True, the use of this machinery by the control-system engineer was often aimed in a direction different from that which might seem most satisfying to a physiologist (e.g., the engineer is vitally concerned with stability because he does not want to build an unstable system; the physiologist is confronted by an existing system whose stability is usually obvious), but this is a minor problem. However, when we looked at even a relatively simple biological system we discovered it to be nonlinear, so that these analytical tools were no longer applicable. Of course the general quantitative conceptual approach and the feedback principle were still just as valuable, and modern computer technology allowed us to obtain particular solutions for the nonlinear equations which were derived. Nevertheless, there is a certain feeling of disappointment in not finding a sufficiently general theory of nonlinear systems available for our use. The nature of the beast is

* L. Stark and P. M. Sherman, "A servoanalytic study of consensual pupil reflex to light," *J. Neurophysiol.* 20 (1957): 17; M. Clynes, "Computer analysis of reflex control and organization: respiratory sinus arrythmia," *Science* 131 (1960): 300; V. W. Bolie, "Coefficients of normal blood glucose regulation," *J. Appl. Physiol.* 16 (1961): 783; J. D. Hardy, "Physiology of temperature regulation," *Physiol. Rev.* 41 (1961): 521; F. E. Yates and J. Urquhart, "Control of plasma concentration of adrenocortical hormones," *Physiol. Rev.* 42 (1962): 359.

such that we may have to relax our expectations of ever finding such a theory. Whereas a single transfer function characterizes the entire behavior of a linear system (i.e., specifies its response to any arbitrary input), a nonlinear system presents a whole new problem whenever the form or even the amplitude of forcing is changed, or whenever initial conditions change.

There are some analytical and graphic methods which can be applied to the analysis of certain nonlinear systems.* One is the "describing function" technique which represents an attempt to extend the methods of frequency analysis considered in Chapter 5 to nonlinear systems. It assumes that there is only a single nonlinear block included in an otherwise linear system and that the input to this block is a pure sinusoid. This amounts to assuming that the linear portion of the system filters out all higher harmonics from the output of the nonlinear block, so that the input to the latter can indeed be a pure sinusoid. Under these conditions, the fundamental term in the Fourier expansion of the nonlinear block output need be the only one considered, and it is used to define an approximate frequency transfer function (or describing function) for this block. Nyquist-type stability tests may be applied and closed-loop frequency response curves obtained. However, the method yields no information about transient response, and cannot be used if more than one nonlinearity is present.

Another method is phase plane analysis, a graphic technique which gives information about transient behavior. Although theoretically applicable to systems of any order, it is practical only for second-order systems. It does not resemble any of the linear techniques but involves plotting (for a second-order system) a family of curves relating $\dot{y}$ and y for each of a large number of sets of initial conditions. The $(y, \dot{y})$ coordinate system is called the phase space, each curve is called a phase trajectory, and the family of curves is the phase portrait. From the latter, a variety of information about the transient response can be derived.

---

* J. G. Truxal, *Automatic Feedback Control System Synthesis* (New York, McGraw-Hill Book Company, Inc., 1955); E. Mishkin and L. Braun, Jr., eds., *Adaptive Control Systems* (New York, McGraw-Hill Book Company, Inc., 1961); G. J. Thaler and M. P. Pastel, *Analysis and Design of Nonlinear Feedback Control Systems* (New York, McGraw-Hill Book Company, Inc., 1962).

Still another method employs random inputs and statistical describing functions. However, if the system contains several nonlinearities or if it is greater than second order (thus including most biological systems), none of these methods is satisfactory and one must turn to computer simulation as we have done.

It is thus apparent that the modern computer (analog or digital) is by all odds the most important new tool available for solution of the large sets of nonlinear equations which will undoubtedly result from the application of quantitative techniques to the analysis of biological control systems. The computer makes possible the solution of mathematical problems which would not even have been attempted just a few short years ago. Thus the modern biologist can hope not only to formulate realistic mathematical models of biological systems, but to solve them (and hence test them) as well.

In a recent symposium, Richard Bellman discussed the role of mathematics and mathematicians in biomedical research.* He pointed out that there are old and new techniques in mathematics as well as old and new problems in biology, and that all four types of application (old techniques to old problems, old techniques to new problems, new techniques to old problems, and new techniques to new problems) probably had something of value to offer to biology. I could not agree more and I do not think it is too early to begin. After all, it is 1963!

* R. Bellman, "Mathematical experimentation and biological research," *Federation Proc.* 21 (1962): 109.

Index